RELATIVISM REFUTED

VOLUME 189

HARVEY SIEGEL
Department of Philosophy, University of Miami

RELATIVISM REFUTED

A Critique of Contemporary Epistemological Relativism

D. REIDEL PUBLISHING COMPANY
A MEMBER OF THE KLUWER ACADEMIC PUBLISHERS GROUP
DORDRECHT / BOSTON / LANCASTER / TOKYO

Library of Congress Cataloging-in-Publication Data

Siegel, Harvey, 1952 –
Relativism refuted.

(Synthese library; v. 189)
Bibliography: p.
Includes index.
1. Relativity. I. Title.
BD221.S58 1987 121 87–15154
ISBN 90–277–2469–5

Published by D. Reidel Publishing Company,
P.O. Box 17, 3300 AA Dordrecht, Holland.

Sold and distributed in the U.S.A. and Canada
by Kluwer Academic Publishers,
101 Philip Drive, Norwell, MA 02061, U.S.A.

In all other countries, sold and distributed
by Kluwer Academic Publishers Group,
P.O. Box 322, 3300 AH Dordrecht, Holland.

Printed in The Netherlands

TABLE OF CONTENTS

For my parents,
Gertrude and Max Siegel

ACKNOWLEDGEMENTS

Several of the chapters of this book have first appeared in the form of journal articles. In the present work each has been at least to some small degree updated, and the "connective tissue" required to render the whole a coherent and cohesive argument supplied. I am grateful to the editors of the various journals for their permission to reprint portions of the original articles here.

Chapter 1 appeared as 'Relativism, Truth and Incoherence', *Synthese* **68**, 1986, pp. 225–259. (D. Reidel Publishing Company).

Chapter 3 appeared as 'Objectivity, Rationality, Incommensurability, and More', *The British Journal for the Philosophy of Science* **31**, 1980, pp. 359–375. The first section of the chapter is taken from my doctoral dissertation, *Kuhn's Philosophy of Science and Science Education* (Harvard University, 1977).

Chapter 4 is taken from two articles: 'Meiland on Scheffler, Kuhn, and Objectivity in Science', *Philosophy of Science* **43**, 1976, pp. 441–448; and 'Epistemological Relativism In Its Latest Form', *Inquiry* **23**, 1980, 107–117. (The Norwegian University Press).

Chapter 5 appeared as 'Brown on Epistemology and the New Philosophy of Science', *Synthese* **56**, 1983. pp. 61–84. (D. Reidel Publishing Company).

Chapter 6 appeared as 'Truth, Problem-Solving and the Rationality of Science', *Studies in History and Philosophy of Science* **14**, 1983, pp. 89–112.

Chapter 7 appeared as 'Goodmanian Relativism', *The Monist* **67**, 1984, pp. 359–375.

PREFACE

I have long been struck by the failure of some others to be struck by the power and cogency of the several self-refutation or incoherence arguments against (epistemological) relativism. Even so, many would question the need for this book. For while many remain unmoved by the arguments for relativism's incoherence, to others relativism is an obvious non-starter, a position whose very statement requires logical contradiction and so self-refutation, and so a position not worthy of sustained philosophical attention.

For a time I shared this latter attitude. But it became clear that such an out-of-hand rejection of relativism was too cavalier. For one thing, many highly regarded philosophers, working in a variety of areas, embraced one or another form of relativism. For another, several philosophers took on the task of articulating and defending relativism, paying special attention to the problem of relativism's alleged incoherence. Most importantly, the last two-three decades have witnessed a striking emergence of relativism in the philosophy of science. These several developments seemed to provide ample reason for a systematic reconsideration of the case for (and against) relativism. Hence this volume.

As an articulated epistemological position, relativism is as old as the Greeks. So is the classical argument alleging relativism's incoherence. It seemed appropriate, therefore, to begin at the beginning by re-examining the argument for relativism put forward by Protagoras in Plato's *Theaetetus*, as well as Socrates' articulation of the argument for relativism's incoherence which also appears in that dialogue. This I do, in the early stages of Chapter 1. But the discussion quickly moves to the contemporary philosophical scene, and the vast majority of the book treats of relativism as it manifests itself in current epistemology and philosophy of science. The main figures here include many of the most respected philosophers of the present, as well as many more who, while less well known, have made important contributions to the contemporary debate. Nevertheless, the presences of Protagoras and Plato are never far removed from the discussion.

The book, in its various segments, has been developing over several years, and I have incurred the intellectual debt of many. Harold I. Brown, Nicholas Burbules, Gerald Doppelt, Catherine Z. Elgin, Edward Erwin, Hartry Field, Alastair Hannay, Sophie Haroutunian-Gordon, Geoffrey Hellman, Carl R. Kordig, Ted Lockhart, Roy Mash, Ernan McMullin, Edward Mooney, Denis Phillips, Donald Provence, Israel Scheffler, Frederick F. Schmitt, Bill Sewell, Anthony Stigliano, Bruce Suttle, and Joseph Ullian have all helpfully commented on at least one chapter, or paper on which a chapter is based, and frequently on several. Max Black, Denis Phillips, and Israel Scheffler deserve special mention for their thoughtful guidance on matters both general and specific. It is a pleasure to acknowledge the help and friendship of these philosophers here. I am grateful as well to my philosophical colleagues at the University of Miami for their support and encouragement, and to that University for its support in the form of a Max Orowitz Summer Research Grant and for the provision of first rate word processing facilities.

INTRODUCTION

Is knowledge (and/or truth) relative – to time, to place, to society, to culture, to historical epoch, to conceptual framework, to personal training or conviction – so that what counts as knowledge, or as truth, depends upon the value of one or more of these variables? Or is knowledge "absolute", or universal, in the sense that they are such independently of these qualifying considerations? If the former, isn't the very idea of knowledge undermined? If the latter, how is this absolutist position to be understood? Does it entail that knowledge requires certainty? Or that there is some privileged position or framework in terms of which all knowledge must be couched? What would such a framework be like?

These questions are as old as philosophy itself. But they have been raised anew by recent developments in epistemology and philosophy of science. This recent work suggests, with dramatic force, that an absolutist stance is untenable, and that some form of relativism must be correct. Yet such a position flies in the face of the widely accepted view that relativism cannot be correct, for it succumbs to a self-destructive, self-referential incoherence. A variety of arguments which purportedly yield the relativistic conclusion, as well as arguments for relativism's incoherence and other arguments which deny the viability of relativism, form the foci of the present study. It will be argued in what follows that the various arguments in support of relativism do not succeed, and that relativism cannot be coherently maintained. A brief sketch of a non-relativistic epistemology, which incorporates the strengths of the newly articulated relativist positions while managing to defeat or avoid the criticisms of "absolutism" made by the "new relativists", and which highlights the central role of an account of rationality in a full account of non-relativist epistemology, will also be presented.

The argument begins, in Part I, Chapter 1, with a discussion of an ancient articulation, defense and criticism of relativism. In the *Theaetetus*, Protagoras and Plato respectively defend and challenge an extreme version of relativism. Their exchange is here analyzed in some detail. Protagoras' position, and Plato's criticisms of it, are re-fashioned in more modern notation. It is argued that these modernized versions of Plato's arguments for relativism's incoherence are successful, and that contemporary relativists have thus far failed to overcome the incoherence charge against their view.

The notion of relative truth looms large in this discussion, and detailed consideration is given to the sustained attempt of the contemporary exponent of relativism, Jack W. Meiland, to overcome the incoherence charge and to develop a relativist position which depends on the notion of relative truth.[1] But attention is also paid to Hartry Field's recent attempt to develop a relativist view which eschews relative truth.[2] It turns out that the latter sort of relativism is as much open to the incoherence charge as its relative-truth cousin.

Chapter 2 centers on the notions of conceptual framework and conceptual scheme. Several philosophers have suggested that the unavoidability of frameworks, and the inescapability of conceptual schemes, necessitates some form of relativism. Wittgenstein's notion of "forms of life" is often said to play a similar role in his philosophy. But other philosophers have been unimpressed by this road to relativism. Karl Popper, for example, has denied the necessity of, and the existence of, an over-arching conceptual framework, and castigates those who affirm such existence as proponents of the "myth of the framework."[3] Donald Davidson[4] has challenged the very meaningfulness of the notion of conceptual scheme. In this chapter the positions of Popper, Davidson, and others are examined, and the relevant literature on the notions of conceptual scheme and framework assessed, the aim being a determination of the extent (if any) to which considerations of frameworks, schemes and the like support a relativist epistemology. It is argued that Popper's criticism of the myth of the framework, while important in certain respects, is not sufficient to blunt the relativist argument, and that Davidson's criticism of the notion of conceptual scheme similarly fails decisively to refute relativism. However, an alternative argument, borrowing mainly from the incoherence argument developed in Chapter 1 but also in part from W.V. Quine,[5] is presented, which suggests that considerations of conceptual schemes and frameworks do not offer in the end any comfort or aid to the relativist.

While relativism is now being advanced from many quarters, and for a wide variety of reasons, there is no doubt that the major spur responsible for the advance of relativism is that provided by recent work in the philosophy of science. Part II is devoted to the connection between relativism and the "new philosophy of science."

It is widely known that the philosophy of science has undergone dramatic changes in the last thirty years. Central tenets of the "old" philosophy of science, i.e., logical positivism or logical empiricism, have come under heavy and sustained attack. Generally rejected is the positivist distinction between theoretical and observation terms.[6] Rejected in many quarters is the important positivist distinction between the contexts of discovery and justification.[7] So too is the positivist view that scientific knowledge is cumulative. The sharp separation between history of science and philosophy of science advocated by the positivists has come under intense criticism, as has the positivist penchant for reconstructing science by way of the logic of the predicate calculus.[8] These criticisms of positivism have gone hand in hand with the effort to construct a post-positivist philosophy of science. This effort is continuing, and at present there is no consensus in philosophy of science as to the proper successor to positivism. Nevertheless, several post-positivist philosophers of science offer views which emphasize the relativity of scientific knowledge, and some argue that the rejection of positivism leads inexorably to a relativist view. In Part II several post-positivist defenders of relativism, as well as some "closet" relativists, who criticize relativism and yet unwittingly offer relativistic views of scientific knowledge, are considered.

The major figure in the new philosophy of science is Thomas S. Kuhn. Kuhn's enormously influential *The Structure of Scientific Revolutions*[9] offers both a major criticism of positivist philosophy of science, and a significant constructive presentation of a post-positivist view. The book has aroused great controversy, not a small portion of which focuses on the alleged relativism of Kuhn's position. But Kuhn has denied being a relativist, and in several later papers he has sought to clarify his views concerning relativism and related issues. Chapter 3 presents a detailed examination of the complex development of Kuhn's position, and of the intricacies of the debate between Kuhn and his critics regarding Kuhnian relativism and more general issues concerning the objectivity and rationality of scientific knowledge.

Several thinkers, in reaction to the revolt against positivism, have sought to defend and develop Kuhn's analysis of science, including his version of relativism. In Chapters 4 and 5 a systematic survey and critique of these supporters of Kuhn is offered. Kuhnians such as Jack W. Meiland,[10] Gerald Doppelt,[11] Harold I. Brown,[12] Wolfgang Stegmüller,[13] and others are considered.

Of course not all participants in the debate concerning the new philosophy of science are defenders of Kuhn. Stephen Toulmin[14] and Larry Laudan,[15] in particular, offer sharp criticisms of Kuhn, and in particular of Kuhnian relativism. Toulmin and Laudan offer their own post-positivist analyses of science. They are both particularly interested in the problem of accounting for the rationality of science. In the process of articulating their respective accounts of science and its rationality, however, both Toulmin and Laudan, despite their explicit desire not to do so, portray scientific knowledge relativistically. In Chapter 6 these "un-Kuhnian" post-positivist analyses of science, and their unwitting acceptance of relativism, are evaluated.

It must be noted that for all the thinkers to be considered, relativism is intertwined with their general analyses of science, and so with a variety of additional issues. It is impossible, for example, to come to grips with Kuhn's treatment of relativism (Chapter 3) without seeing how that treatment meshes with his discussion of rationality, incommensurability, and a host of other notions and theses. The same is true for the Kuhnians to be discussed in Chapters 4 and 5, and for Toulmin and Laudan, to be discussed in Chapter 6. Consequently, what is presented in Part II is not only a consideration of relativism; it constitutes rather a general critique of several important formulations of the new philosophy of science, taken as wholes. While relativism is the thread, the analyses offered go beyond a concern with relativism alone. Because relativism is intertwined with other issues central to the new philosophy of science (in particular those concerning the rationality and objectivity of science), such a general critique as is offered is in fact the only sort possible. In any case, relativism surfaces frequently enough to be a thick thread, and to hold together the varied separate discussions which make up Part II.

It is hoped that our understanding of the issues discussed in Part II, which are central to major articulations of the new philosophy of science, will benefit from being studied in the more general epistemological context provided by the book as a whole.

Part III departs from philosophy of science and takes up additional epistemological topics relevant to relativism.

Chapter 7 focuses on the work of Nelson Goodman. Goodman has long proclaimed himself a relativist, although his earlier work[16] may with justice be regarded as pluralist rather than relativist. In his most recent work[17] Goodman declares his view "radically relativist" (albeit within "rigorous restraints"), but these restraints, which manifest themselves mainly in terms of "standards of rightness", make his overall commitment to relativism problematic. It is argued in this chapter that there are actually two versions of "Goodmanian Relativism", that the first is defensible but is not truly relativist, and that the second is a genuinely relativist position but is not defensible.

In Chapter 8 an "absolutist" epistemology is sketched. It is argued that a rejection of relativism need not entail an epistemology which embraces infallibility, certainty, unrevisability, dogmatism, or any particular privileged framework. A fallibilist, revisable, realist absolutism, it is argued, is both coherent (contrary to the charges of some relativists) and free of the flaws often attributed to "vulgar" absolutism. This chapter also outlines the direction of further work. First, there are several domains in which relativism still needs to be considered: in the philosophy of the social sciences, especially the literature spawned by Wittgenstein and Winch; in the sociology of knowledge, and in particular the sociology of science, as developed by contemporary figures such as Michael Young, David Bloor, and Barry Barnes; and in ethics, where moral relativism has been revived recently by Williams, Mackie and Harman. Second, there is a deep connection between the issues of relativism and rationality, in that relativism threatens to undermine the very possibility of rationality. Consequently, it is imperative that the implications for rationality of the book's anti-relativist arguments be spelled out, and, more importantly, that a non-relativist analysis and defense of the ideal of rationality be worked out. This latter should involve consideration of the role of reason in ethics, of the rationality of science, of the problems for the notion of rationality raised by puzzles like the prisoner's dilemma, and so on. All of this constitutes the basis of future research, and the subject matter of a sequel to the present volume.

Despite the sketch of an absolutist epistemology offered in Chapter 8, the book as a whole is critical rather than constructive. Given the popularity and influence of relativism in contemporary philosophy (and beyond), this seems appropriate. The constructive tasks of developing a full fledged absolutist epistemology, and a theory of rationality necessary for that epistemology, are beyond the scope of the present work and are perforce left for the sequel.

PART I

RELATIVISM AND THE PROBLEM OF INCOHERENCE

CHAPTER 1

THE INCOHERENCE ARGUMENT AND THE NOTION OF RELATIVE TRUTH

Epistemological relativism has been defended by a variety of thinkers stretching back at least as far as Protagoras. For just as long, however, others have thought the doctrine to be incoherent because self-refuting. This is perhaps the most fundamental challenge faced by the relativist. The present investigation begins, therefore, by considering the status of the incoherence charge.

In this chapter I review the debate between Plato and Protagoras regarding relativism and incoherence, and consider a series of more recent defenses of epistemological relativism. A central focus will be the role that the notion of relative truth plays in the various arguments for relativism, both ancient and contemporary; though a defense of relativism that eschews that notion, developed by Hartry Field, will also be examined. I shall argue that, like Protagoras, the recent defenders fail to meet the challenge posed by the incoherence charge, and that the doctrine of epistemological relativism remains untenable because incoherent.

PROTAGOREAN RELATIVISM AND THE SOCRATIC ARGUMENTS FOR INCOHERENCE

In the *Theaetetus*, Protagoras is portrayed as holding that "man is the measure of all things," and that any given thing "is to me such as it appears to me, and is to you such as it appears to you."[1] In considering Theaetetus' suggestion that knowledge is perception, Socrates concludes that it is equivalent to Protagoras' view: "Then my perception is true for me, for its object at any moment is my reality, and I am, as Protagoras says, a judge of what is for me, that it is, and of what is not, that it is not."[2] Protagoras holds, according to Socrates, that "Each one of us is a measure of what is and of what is not. . . . To the sick man his food appears sour and is so; to the healthy man it is and appears the opposite. Now there is no call to represent either of the two as wiser – that cannot be – nor is the sick man to be pronounced unwise because he thinks as he does, or the healthy man wise because he thinks differently. . . . In this way it is true . . . that no one thinks falsely."[3] Socrates encapsulates Protagoras' relativism as consisting in the view that "what seems true to anyone is true for him to whom it seems so."[4]

Protagoras' view is an extreme version of relativism: knowledge and truth are relative to the person contemplating the proposition in question. *p* is true (for me) if it so seems; false (for me)[5] if it so seems. Since the final arbiter of truth and knowledge is the individual, Protagoras' view denies the existence of any standard or criterion higher than the individual by which claims to truth and knowledge can be adjudicated.

Socrates offers several arguments against the Protagorean view. Two in particular will be of interest here.[6] The first questions the justifiability of Protagoras' sophistical activity, given his doctrine:

> If what every man believes as a result of perception is indeed to be true for him; if, just as no one is to be a better judge of what another experiences, so no one is better entitled to consider whether what another thinks is true or false, and . . . every man is to have his own beliefs for himself alone and they are all right and true – then . . . where is the wisdom of Protagoras, to justify his setting up to teach others and to be handsomely paid for it, and where is our comparative ignorance or the need for us to go and sit at his feet, when each of us is himself the measure of his own wisdom? . . . to set about overhauling and testing one another's notions and opinions when those of each and every one are right, is a tedious and monstrous display of folly, if the Truth of Protagoras is really truthful. . . .[7]

Here Socrates levels the first version of the incoherence charge: Protagoras is involved in the project of "overhauling and testing one another's notions and opinions" – that is, he is engaged in the epistemological task of assessing the warrant and justification of knowledge-claims – but his thesis undermines that very project, since if his thesis is right, then there is no chance of any thesis failing a test of adequacy, or being judged unjustified or unwarranted, because the rival theses "of each and every one are right." If knowledge is relative, then the task of judging claims to knowledge is pointless. If Protagoras' thesis is right, it cannot be right, for it undermines the very notion of rightness. Protagorean relativism is thus self-defeating – if it is right, it cannot be right – and so is incoherent. Let us call this first argument for the incoherence charge the 'undermines the very notion of rightness' (henceforth *UVNR*) argument: relativism is incoherent because, if it is right, the very notion of rightness is undermined, in which case relativism cannot be right.

The second argument for the incoherence charge focuses on the

Protagorean view that all opinions are true for those who believe them, and concomitantly that no sincerely held opinion is false ("that no one thinks falsely"). Socrates argues that this thesis cannot be correct, and in fact is self-defeating, for if true, some beliefs will be false – contrary to the thesis. Socrates' argument[8] centers on the phenomenon of conflicting opinion. Suppose A believes p, and B believes not-p. p is true for A, according to Protagoras, yet false for B. Then p is true for some, and false for others. Now, suppose that p is a statement expressing Protagorean relativism. Then Protagorean relativism is false for all those who do not believe it: it is false for all if no one believes it; and true only to the extent that some number of people (perhaps only Protagoras himself) believe it:

> Supposing that not even he [i.e. Protagoras] believed in man being the measure and the world in general did not believe it either – as in fact it doesn't – then this *Truth* which he wrote would not be true for anyone. If, on the other hand, he did believe it, but the mass of mankind does not agree with him, then, you see, it is more false than true by just so much as the unbelievers outnumber the believers.[9]

In fact, Socrates argues, the situation is even worse for Protagoras than that. For Protagoras seems to have to acknowledge, given his doctrine, that his opponents' view that he is wrong is itself right:

> [Moreover,] Protagoras, for his part, admitting as he does that everybody's opinion is true, must acknowledge the truth of his opponents' belief about his own belief, where they think he is wrong.[10]

Protagoras would be forced, Socrates argues, to:

> acknowledge his own belief to be false, if he admits that the belief of those who think him wrong is true . . . [for he] admits that this opinion of theirs is as true as any other.[11]

So, Socrates argues, Protagoras is bound by his own lights to grant the truth of his opponents' beliefs, even in the case where their belief is that Protagorean relativism is false. And since their opinion is true, according to the Protagorean doctrine, and their opinion is that that doctrine is false, then that doctrine is false – even for Protagoras himself.[12] Thus Protagorean relativism is self-defeating in a second way. If opinions conflict, and the doctrine holds that all opinions are true, then some opinions cannot be true. In particular, if opinions conflict about the truth of Protagorean relativism, then the Protagorean relativist must acknowledge the truth of the opinion that that doctrine is false. Thus, if it is true, then (as long as there is one who holds that it is false) it is false.

Hence the doctrine is self-defeating, and so incoherent. Let us call this second argument for the incoherence charge the 'necessarily some beliefs are false' (henceforth *NSBF*) argument: relativism is incoherent because it holds that all beliefs and opinions are true, yet, given conflicting beliefs, some beliefs must necessarily be false – in which case relativism cannot be true.

We have seen thus far two independent arguments for the incoherence of Protagorean relativism.[13] The *UVNR* argument concludes that such relativism is incoherent because, if right, its rightness cannot be established, because the very notion of rightness is undermined – so that, if right, it cannot be right. The *NSBF* argument concludes that Protagorean relativism is incoherent because, if true, then it is false (so long as at least one person is of the opinion that it is false), because the Protagorean relativist is bound by her doctrine to regard all opinions as true, including the opinion that that doctrine is false. Our question is: Are these arguments conclusive? Is Protagorean relativism incoherent? Several recent writers have argued that it is not; their defenses of the doctrine, or of some related version of epistemological relativism, will be the focus of attention during most of the remainder of this chapter. First, however, it will be helpful to consider more contemporary, and more generalized, versions of the Protagorean doctrine and the Socratic arguments for its incoherence.

EPISTEMOLOGICAL RELATIVISM AND THE INCOHERENCE CHARGE IN THEIR MODERN DRESSES

Epistemological relativism may, without violence to Protagoras or more recent defenders, be characterized as follows:[14]

ER: For any knowledge-claim p, p can be evaluated (assessed, established, etc.) only according to (with reference to) one or another set of background principles and standards of evaluation $s_1, \ldots s_n$; and, given a different set (or sets) of background principles and standards $s'_1, \ldots s'_n$, there is no neutral (that is, neutral with respect to the two (or more) alternative sets of principles and standards) way of choosing between the two (or more) alternative sets in evaluating p with respect to truth or rational justification. p's truth and rational justifiability are relative to the standards used in evaluating p.

The Protagorean relativist will assent to *ER*, since it captures the intuition that knowledge and truth are relative to each individual thinker: if p is true (or a genuine item of knowledge) according to my standards, then it is true (for me), and there is no other standard by which my individual judgments and standards may themselves be neutrally evaluated. *ER* is more general than Protagorean relativism, however, for it places the source of relativism at the level of standards rather than the level of personal opinion or perception, and as such aptly characterizes more recent relativisms such as those of Kuhn, Young, Bloor, Barnes, Winch, Wittgenstein, and others.[15]

It immediately follows from *ER* that if p is evaluated differently according to two different sets of principles and standards – that is, for example, if p is true (probably true, highly confirmed, etc.) according to $s_1, \ldots s_n$, but false (probably false, poorly confirmed, etc.) according to $s'_1, \ldots s'_n$, there is no way to evaluate these conflicting evaluations. p is true relative to $s_1, \ldots s_n$, and false relative to $s'_1, \ldots s'_n$. Since there is no neutral way of evaluating the rival sets of standards, there is no neutral way of evaluating conflicting evaluations of p. Such evaluations can only be carried out relative to one or another set of principles and criteria of evaluation. Given my set of standards, opinions, and convictions, p's epistemic status is as I judge it to be.

If *ER* effectively characterizes epistemological relativism, how can we reformulate, with respect to *ER*, Socrates' arguments against Protagorean relativism? The second (*NSBF*) argument is easily recast. Instantiating *ER* in the original formulation yields:

ER': *ER* can be evaluated (assessed, established, etc.) only according to (with reference to) one or another set of background principles and standards of evaluation $s_1, \ldots s_n$; and, given a different set (or sets) of background principles and standards of evaluation $s'_1, \ldots s'_n$, there is no neutral (that is, neutral with respect to the two (or more) alternative sets of principles and standards) way of choosing between the two (or more) alternative sets in evaluating *ER* with respect to truth or rational justification. *ER*'s truth and rational justifiability are relative to the standards utilized in evaluating *ER*.

If *ER* is true, then, as *ER'* states, *ER* is itself relative to alternative, and equally legitimate, sets of background principles and standards of evaluation. Since these alternative sets will suggest differing evaluations of *ER*, and since there is no way neutrally to pick one evaluation over and against any others, it follows that, if *ER* is true, then *ER*'s truth will vary according to the principles and criteria by which *ER* is evaluated. In particular, it follows that, if according to some set of standards $s_1, \ldots s_n$ *ER* is judged to be false, then, if *ER* is true, (at least according to that set of standards $s_1, \ldots s_n$) *ER* is false.[16] (This neatly mirrors the Socratic argument according to which Protagoras is bound by his own principles to recognize the falsity of Protagorean relativism, so long as someone is of the opinion that it is false.) In this way, *ER* is self-refuting, and so incoherent.

Socrates' first (UVNR) argument has it that Protagorean relativism is self-refuting in that, if it is right, it undermines the very idea of rightness, and so cannot be right. Another way to put this point is that the notion of rightness is intelligible only when understood nonrelativistically. "Relative rightness" is not rightness at all. For the relativist wants to argue that relativism is right (or true, or cognitively superior) and that non-relativism is wrong (or false, or cognitively inferior), or less adequate philosophically, than relativism. To make this claim non-relativistically, however, is to give up relativism; conversely, to make the claim only relatively is not to make it at all.

Put in terms of *ER*, the *UVNR* argument can be cast as follows. Assume *ER* to be a rationally justifiable position. Then there are good reasons for holding *ER*. But good reasons cannot be biased or non-neutral or arbitrary or idiosyncratic (by definition of 'good reason' – this point will be pursued further below). Therefore, if *ER* is rationally justifiable, there must be some non-relative, neutral (with respect to the presuppositions of relativists and non-relativists) framework or ground from which we can make that judgment. Thus *ER*, which denies the possibility of such a framework, is incorrect. In short, if relativism is rationally justifiable, it must have a non-relativistic ground, which possibility it denies. Thus *ER*, if true, is not rationally justifiable, since if *ER* is true there can be no neutral ground from which to assess the rational justifiability of any claim, including *ER* itself. Moreover, if *ER* is (true and) rationally justifiable, then it is false, for the rational defense of *ER* requires the sort of non-relativistic ground which *ER* itself denies. Thus *ER* is either not rationally justifiable, or false. The assertion and defense of *ER* is thus self-refuting, and so incoherent.

This argument points out what Socrates' first argument points out: namely, that the relativist must appeal to non-relativistic criteria, and assert relativism non-relativistically, in order to make the case for relativism. This is self-defeating for the relativist. But to fail to assert and defend relativism in this (non-relativistic) way is to fail to join the issue with the non-relativist who asserts that relativism is false (or incoherent). So the relativist can defend relativism only by rendering it incoherent. Conversely, to defend relativism relativistically is to fail to defend it at all. For if relativism is right, the very notion of rightness, and indeed that of rational defense, is given up, and so it cannot coherently be claimed that relativism is right or rationally defensible. In short: to defend relativism is to defend it non-relativistically, which is to give it up; to 'defend' it relativistically is not to *defend* it at all. And this is precisely the lesson of Socrates' first argument.[17]

DO THE INCOHERENCE ARGUMENTS BEG THE QUESTION? A RELATIVIST CONCEPTION OF TRUTH

Several recent writers have sought to defend relativism from the charge of incoherence. The first major line of defense to be considered is that which holds that the incoherence arguments beg the question by assuming an absolutist conception of knowledge or truth.

Harold I. Brown, for example, writes, with respect to the *NSBF* argument, that

> . . . this argument has no force against any consistent relativism, even the extreme relativism of Protagoras. Its apparent cogency derives from a tacit acceptance of the absolutist assumption that we are justified in making a knowledge claim only if it is based on an unquestionable foundation. Given this assumption, once we admit the possibility of knowledge claims contrary to but as well founded as our own, our own knowledge claims become illegitimate. But the acceptability of this thesis is the central issue in dispute between relativism and absolutism.[18]

Thus, Brown concludes, the absolutist begs the question against the relativist by assuming absolutism.

The problem with Brown's argument against the absolutist is that it confuses absolutism with foundationalism. If relativism is the view that knowledge and truth are relative to framework, conceptual scheme, paradigm, cultures, personal predilection, etc., and there are no criteria or standards by which claims put forth by rival positions can be fairly, neutrally, or objectively judged, then absolutism should be understood as the (contrary) view that such claims can be evaluated in a non-question-begging way, and that objective comparison of rival claims is possible. But absolutism so construed is not at all tantamount to foundationalism, that is, to the epistemological thesis that "knowledge requires an indubitable foundation and that knowledge is developed by building on that foundation."[19] One can hold that knowledge is absolute in the sense that claims to knowledge can be fairly, non-question-beggingly assessed without holding further that knowledge "requires an indubitable foundation." Brown here confuses absolutism with foundationalism; he similarly confuses relativism with *fallibilism*:

The main thesis of relativist epistemology is that knowledge can be constructed on a fallible foundation. Relativism affirms my right to hold my own presuppositions in spite of their fallibility, to proceed on the basis of these presuppositions, and to reject competing sets of presuppositions as false.[20]

But these rights are affirmed, not by relativism, but by fallibilism. One needn't be a relativist to affirm the first two rights Brown mentions; moreover, it is not clear (as we shall see below, note 49) that the relativist can consistently affirm the third right. Relativism's commitment to the non-existence of neutral standards aligns it, not with fallibilism, but with *arbitrariness*. It is this arbitrariness which makes relativism the radical and potentially destructive doctrine that it is perceived by its critics to be. In any case, one can consistently espouse a fallibilist absolutism – that is, one can consistently hold that knowledge is fallible, and not certain or indubitable, and also that claims to knowledge can be neutrally and objectively evaluated and assessed.[21]

A more systematic attempt to discredit the incoherence arguments and to develop a positive case for epistemological relativism is that of Jack W. Meiland. In a series of articles,[22] Meiland denies that relativism is self-refuting, and he has developed an analysis of relative truth which, he argues, stands as a legitimate alternative to an absolutist conception of truth. In fact, these two efforts are connected in Meiland's work, for it is by establishing a viable conception of relative truth, according to Meiland, that relativism escapes the self-refutation problem:

That relativism is self-refuting . . . is a myth which must be laid to rest. It *would* be inconsistent for the relativist to say both that all doctrines are relatively true and that relativism is not relatively true but instead is absolutely true. However, the careful relativist would not and need not say this. He would either say that all doctrines except relativism (and perhaps its competitors on the meta-level) are relatively true or false, or else he would say that his own doctrine of relativism is relatively true too. And saying that relativism is only relatively true does not produce inconsistency.[23]

The first alternative Meiland mentions is unhelpful for the relativist, in the absence of an argument which reasonably distinguishes between relativism (and perhaps its competitors on the meta-level) and all other doctrines and claims with respect to truth such that the former, but not the latter, can properly be regarded as either true or false absolutely. Meiland has furnished no such argument; nor is it easy to see on what consideration such an argument might be based.[24] It is Meiland's second alternative – that the self-refutation charge can be avoided by holding that the doctrine of relativism is itself only relatively true – that is worth scrutiny here. And this alternative clearly rests on the coherence of the notion of relative truth. It is no surprise, therefore, that Meiland seeks to establish the tenability of that notion.

Meiland's discussion of relative truth hinges on his claim that relative truth can be understood independently of the concept of absolute truth – for, as he rightly acknowledges, relativism would not escape the self-refutation problem if it denied the legitimacy of absolute truth by affirming a concept of relative truth which itself depended on or appealed to the absolute conception.[25] Meiland offers an analysis of absolute and relative truth which he suggests is such that the latter does not depend on the former. On this analysis, absolute truth is a two-term relation, while relative truth is a three-term relation:

(1) The concept of absolute truth seems to be a concept of a *two*-term relation between statements (or perhaps propositions) on the one hand and facts (or states of affairs) on the other. But the concept of relative truth, as used by some relativists, seems to be a concept of a *three*-term relation between statements, the world, and a third term which is either persons, world views, or historical and cultural situations.

(2) The relation denoted by the expression 'absolute truth' is often said to be that of correspondence. The relativist can make use of this type of notion and say that "P is true relative to W" means something like "P corresponds to the facts from the point of view of W" (Where W is a person, a set of leading principles, a world view, or a situation).[26]

Meiland acknowledges that the analysis of the concept of relative truth here offered "is very incomplete and raises more questions than it answers."[27] He puts it forward, nonetheless, for the following purpose:

My point in putting this suggestion forward is to begin to show that . . . absolutists are making a great mistake by assuming that relative truth must be either nothing at all or else a variety of absolute truth.[28]

And Meiland summarizes what he takes to be the significance of his analysis as follows:

When we use expressions of the form 'Ø is true for W', it seems legitimate to ask the question "What does 'true' mean in this expression?" . . . The correct relativist answer to this question is: "It means that Ø is true-for-W." The hyphens in this answer are extremely important. For they show that the relativist is not talking about truth but instead about truth-for-W. Thus, one can no more reasonably ask what 'true' means in the expression 'true-for-W' than one can ask what 'cat' means in the word 'cattle'. 'True-for-W' denotes a special three-term relation which does not include the two-term relation of absolute truth as a distinct part.[29]

Meiland offers his analysis for two (related) purposes: he wants to establish the concept of relative truth as a viable, coherent concept which the relativist can appeal to as an alternative to the concept of absolute truth; and he wants to show that relative truth, once shown to be viable, enables the relativist to escape the problem of self-refutation. Unfortunately, neither of these ends are achieved by the analysis offered. I consider them in turn.

THE VIABILITY OF THE CONCEPT OF RELATIVE TRUTH

Consider first Meiland's analysis of relative truth as a "three-term relation between statements, the world, and a third term which is either persons, world views, or historical and cultural situations." To be a genuine three-term relation, it must be possible to individuate each relatum and to distinguish each from the other two. On a conception of absolute truth, this condition (as Meiland recognizes) is easily met: the world, however difficult to apprehend directly, is clearly distinguishable from statements about it. What, however, is the status of the world on the three-term conception? Is it clearly distinguishable from the other two relata? Unfortunately, the answer is no. On the relativist conception, the world is not distinguishable from the third relatum (either persons, world views, or historical and cultural situations). What are related by the alleged three-term relation are statements and the-world-relative-to-W (where W is a person, a set of leading principles, a world view, or a situation – in short, where W is the third relatum). On the relativist conception, the world cannot be conceived as independent of W; if it is so conceived, the relativist conception collapses into an absolutist one, for it is granted that there is a way the world is, independent of statements and of W's. This is precisely what the relativist must deny, however.[30] So Meiland's three-term relation collapses into a two-term relation, between statements and the-world-relative-to-W, or, in a Goodman's terminology, between statements and world-versions.[31] This point can be seen in Meiland's remarks about the relativist's ability to utilize the notion of relations of correspondence. Meiland writes that the relativist "can make use of this type of notion and say that 'P is true relative to W' means something like 'P corresponds to the facts from the point of view of W'."[32] Grant Meiland the use of relations of correspondence: what corresponds is not statements, independent facts, and some W; but rather statements and facts-from-the-point-of-view-of-W. Thus Meiland's three-term relation turns out upon inspection to be a two-term relation.

Now this point by itself is not very damaging to Meiland's position. He can still argue that the two-term relativist conception of truth is distinct from the two-term absolutist conception – the former relates statements and facts-relative-to-W, while the latter relates statements and facts (which are not relative to any W) – and that the relativist conception does not appeal to, incorporate, or rely upon the absolutist conception. And this last point is the point Meiland is most concerned to make: "'True-for-W' denotes a special . . . relation which does not include the . . . relation of absolute truth as a distinct part."[33] This point is important for Meiland because, if it can be sustained, it rescues the relativist from the charge that she relies on the notion of absolute truth in holding a concept of relative truth, thereby refuting her own position by relying on a concept she expressly rejects. Is it the case, however, that the relativist conception does avoid reliance on the absolute conception? It is not clear that it does. Consider Meiland's reasoning here. He writes that the hyphens in 'true-for-W'

> are extremely important. For they show that the relativist is not talking about truth but instead about truth-for-W. Thus, one can no more reasonably ask what 'true' means in the expression 'true-for-W' than one can ask what 'cat' means in the word 'cattle'.[34]

There are two points to make here. The first is that Meiland's conclusion does not follow, for 'true' is not related to 'true-for-W' as 'cat' is related to 'cattle'. 'Cattle' is made up of the concatenation of the letters 'c', 'a', 't', 't', 'l', and 'e'; 'cat' is no more a meaningful part of 'cattle' than is 'ca', 'catt', 'cattl', etc. 'True-for-W', on the other hand, is made up of the hyphenization of distinct concepts which are independently meaningful.[35] The occurrence of 'true' in 'true-for-W' is *not* like the occurrence of 'tru' or 'ru' in 'true-for-W'. In fact, if the role of 'true' in 'true-for-W' is analogous to that of 'cat' in 'cattle', it is difficult to understand why Meiland would bother with his discussion. For it is, after all, a conception of relative *truth* he is concerned to articulate. The hyphens in 'true-for-W' show that on the relativist conception truth is always to be regarded as relative to some W; that it is never to be conceived of as independent of every W. But it does not follow that the hyphenated phrase "does not include the concept of absolute truth as a distinct part."[36] If this did follow, it would be difficult to see why the hyphenated phrase constituted a conception of relative *truth* at all. Adding the hyphens does not eliminate the concept of truth from the hyphenated phrase – as well it shouldn't, if the aim of the relativist's project is to articulate a conception of truth. 'True' in 'true-for-W' is not analogous to 'cat' in 'cattle'; a better analogy is with 'action' in 'action-at-a-distance'. The latter phrase is to be distinguished from contiguous, mechanically connected action; it is still a conception, nonetheless, of *action*. Similarly, 'true-for-W' is to be distinguished from absolute truth, i.e. truth that is not relative to any W; it is still a conception, nonetheless, of *truth*.[37]

The immediate upshot of the above discussion is that Meiland has not shown that his conception of relative truth avoids reliance on the absolute conception, and so he has not shown that his analysis avoids the self-refutation problem it is designed to avoid. There is a further point to note. Meiland is out to establish the coherence and tenability of the notion of relative truth. Grant for the sake of argument what I have just argued should not to be granted, namely that 'true-for-W' does not include the concept of absolute truth as a distinct part. What, then, does 'true-for-W' mean? If it has nothing to do with the concept of (absolute) truth, what does it have to do with? If we grant Meiland his point, all that follows is that he has developed a string of symbols which is not to be understood as containing the concept of absolute truth as a distinct part. He has provided no hint about what the string does mean or refer to. Consequently, his analysis, even if accepted, does not help to establish the meaningfulness, coherence or tenability of the concept of relative truth. For we have only been told what the concept of relative truth is not; we have not been told what it is to be 'true-for-W'.

This latter point deserves additional comment. The task the relativist faces is that of establishing the coherence or tenability of the concept of relative truth. To accomplish the task the relativist must establish a meaning for the concept which does not rely on or collapse into the concept of absolute truth (on pain of the sort of self-refutation Meiland is concerned to avoid in the discussion we have been considering). But the concept of relative truth to be established must also, if it is to accomplish the relativist's task, avoid collapsing into 'mere belief'. That is, 'P is true-for-W' must come to something different than 'W believes that P' (if W is a person), or 'A person in W believes that P' (if W is a historical or cultural situation), or 'A person of W believes that P' (if W is a perspective, conceptual scheme, world view, or set of leading principles), etc. Otherwise all the relativist's talk about relative truth amounts merely to the trivial thesis that persons have different beliefs – hardly a challenging epistemological doctrine, since it in no way precludes the posing of the absolutist question regarding the truth of the various beliefs.[38] In short, the task of the relativist is to establish a meaning for the concept of relative truth according to which 'P is true-for-W' is distinct both from 'W believes that P' and from 'P is (absolutely) true for W'.

Meiland seeks to establish such a meaning in the course of a discussion of the anti-relativist charge that the concept of relative truth cannot be coherently articulated without appealing to an absolutist conception of truth.[39] The charge is easily stated: Relativism holds that a person X's believing that *p* is true entails that *p* is true for X. But, as Meiland articulates the charge:

> When we talk about what Jones believes, we are presumably talking about what Jones believes *to be true* – and, apparently, about what Jones believes to be *absolutely* true. So it appears that his [i.e. Protagoras'] variety of relativism cannot even be stated without recourse to a concept of absolute truth. . . . Relativism is supposed to eschew absolute truth entirely and it allegedly cannot succeed in doing this.[40]

Meiland attempts to circumvent the problem by distinguishing between *p*'s *being* absolutely true and X's *believing* that *p* is absolutely true:

> The relativist can certainly admit that someone *believes* a statement to be absolutely true. The relativist will merely deny that this belief is itself absolutely true (although it may be relatively true – true for the person who believes it) and hence will also deny that the statement which is the object of that belief is itself absolutely true. The important point here is that in stating his position, the Protagorean relativist is not saying that anything is absolutely true; he is only allowing that some people believe that various statements are absolutely true.[41]

The point is well taken. We should not expect the relativist to pretend she has never heard of absolute truth. However, Meiland's response raises a further problem for the relativist, one at least as serious as the one the response is designed to circumvent. As the most recent quotation suggests, all that is required for *p* to be relatively true is that X believe it: X's belief that *p* is true (absolutely or relatively) entails that p is relatively true, i.e. true-for-X. Belief is a sufficient condition for relative truth; for *p* to be relatively true, all that is required is that someone believe it. In short: '*p* is true-for-X' amounts simply to 'X believes that *p*'.

But, as noted above, the relativist project of demonstrating the coherence and tenability of the concept of relative truth is not accomplished by the above analysis. For if relative truth amounts simply to belief, then the relativist thesis is trivialized, and we are left wondering what the fuss is all about. That there exist rival, incompatible truths reduces to the innocuous observation that there exist differences of opinion, i.e. rival, incompatible beliefs. Moreover, if relative truth amounts merely to belief, then the obvious question about the criticizability and justifiability of alternative beliefs is not spoken to; the relativist must be seen as making no claims about the immunity of relative truths – i.e. beliefs – from neutral or objective criticism. But of course a robust and epistemologically interesting relativism must make such claims. In short, epistemological relativism, if it is to be worth attending to, must say more of relative truth than that it comes down to mere belief. Meiland himself, in another context, recognizes the point:

> To say that some belief or practice is true or right for a society or an individual is to say more than simply that that society or that individual holds that belief or practices that practice. If all that was meant by saying a belief is true for Jones is that Jones holds that belief, then every belief that Jones holds would be true for Jones. But the relativist rejects this notion of relative truth; he or she takes the notion of relative truth more seriously than this.[42]

The problem on the table is that Meiland's discussion of the charge that the notion of relative truth cannot be coherently articulated seems to depend upon taking the notion of relative truth in just this unserious way.

Meiland acknowledges the problem; he does not want 'Ø is true for X' simply to mean that X believes that Ø. Rather, he suggests that 'X believes that Ø is true' is to be taken as the *criterion* for saying that Ø is true for X, but that such belief "is not what is meant by saying that Ø is true for him."[43] To the obvious next question, "if 'Ø is true-for-X' does not mean 'X believes that Ø', what does it mean?", Meiland offers the following answer: ". . . the statement Ø is true for Jones *means* 'Ø corresponds to reality for Jones'."[44] But this answer does not help, for 'Ø corresponds to reality for Jones' is opaque unless it means 'Jones believes that Ø corresponds to reality'. In other words, Meiland's answer fails to provide an alternative to the unhelpful view that relative truth amounts merely to belief.

Meiland is aware of the unhelpfulness of the phrase 'corresponds with reality for Jones'. Rather than try to explicate it, or to show that it does not simply collapse into a claim about Jones' belief, however, Meiland contents himself with arguing that the relativist is in no worse a position on this score than the absolutist:

> . . . if the relativist is asked the embarrassing question "what do you mean by 'corresponds with reality for Jones'?", we need only point out that although this question is embarrassing in the sense that it is difficult for the relativist to give any useful answer to it, nevertheless the relativist is in no worse a position than the absolutist at this point. Often, when the absolutist is asked to explain what he means by 'true', he does so in terms of correspondence with reality. And the absolutist who responds in this way owes us an account of what correspondence with reality consists in. Absolutists have not been able to say anything at all satisfactory on this score. Thus, relativism is not to be faulted for being unable to give an account of that which the absolutist cannot give an account of in his own position either.[45]

Meiland is here admitting that his notion of 'correspondence with reality for X' is unhelpful; it does not further his positive account of relativism at all to note that the absolutist faces a parallel problem. But Meiland's situation here is even worse than he realizes, for it is not the case that the absolutist and the relativist are on equally difficult ground here. The absolutist faces the hoary philosophical problem of accounting for a reality which cannot be directly perceived or known. But this is a problem which is independent of the absolutism/relativism controversy. For the purposes of her debate with the relativist, all the absolutist needs is a distinction between reality and her conception of that reality. One can acknowledge that distinction while accepting that reality as such is known to us only through our perceptions, concepts, etc. The relativist, however, needs more than the distinction between reality and our conception of it. The relativist also needs an account of what it is to be 'reality-for-a-person', where that phrase must denote neither reality as such (i.e. independently of any conception of reality) *nor* a person's beliefs concerning or conception of that reality – for the latter alternative reduces, as we have seen, simply to belief. Thus the notion of 'correspondence-with-reality-for-X' is *more* troublesome than the (admittedly troublesome) notion of 'correspondence-with-reality'; it is not the case, as Meiland claims, that the relativist is no worse off than the absolutist here.

The upshot is that Meiland has not shown that his development of the concept of relative truth is able to distinguish that concept from mere belief, so that '*p* is true for X' is anything more than 'X believes that *p*'. Earlier we saw that Meiland also failed to establish that the concept of relative truth does not in the end presuppose or rely on an absolutist conception of truth. And since, for the concept of relative truth to be coherent, tenable, and philosophically interesting, it must have a definite meaning which is distinct from mere belief and which does not presuppose an absolutist conception, we may conclude that Meiland's analysis of the concept of relative truth fails to establish that concept's coherence or tenability.[46]

RELATIVE TRUTH AND SELF-REFUTATION

As noted earlier, one of the classical objections to epistemological relativism is that the concept of relative truth, on which the doctrine of relativism seems to depend, is incoherent (or trivial). Meiland's analysis has given us no reason to doubt the cogency of the classical objection. However, there is still a deeper problem. For even if it could be shown that the concept of relative truth is coherent, that by itself would not establish the cogency of epistemological relativism. I now want to argue that, even if some conception of relative truth were to be shown to be coherent, the doctrine of epistemological relativism would nevertheless remain unworthy of our embrace, for it would still fail to avoid the problem of self-refutation. Specifically, a coherent conception of relative truth fails to enable the relativist to avoid the perils of the *UVNR* argument for the incoherence of relativism discussed above.[47]

Recall that the *UVNR* argument has it that relativism is self-refuting in that, if relativism is right, it undermines the very idea of rightness. For to say that relativism is *right* is to presuppose the existence of nonrelative criteria of rightness by which the judgment that relativism is right can be made. But relativism denies the existence of any such criteria. Thus, if relativism is right, it undermines the very notion of rightness, by denying the very criteria necessary for the judgment of its rightness to have cognitive or epistemic force.

Now, if the relativist embraces the concept of relative truth, she embraces the thesis that any claim is true for those who believe it. (Meiland says, recall, that while 'Ø is true-for-X' does not *mean* 'X believes that Ø', nevertheless such belief is the *criterion* for relative truth.) To embrace this thesis, however, is to accept that the establishment of a claim or proposition or belief *p* as relatively true is not in any way an establishment of the epistemic worthiness of *p*. If *p* is relatively true, according to the relativist, then *p* is believed by the person it is true for; it corresponds with that person's conception of reality. The relative truth of *p* thus renders *p* no more worthy of belief than the equally (relatively) true not-*p* or arbitrary belief *q*. The relativist conception of truth commits one to the view that relative truths are not in any way cognitively superior or preferable to their contradictories, to relative falsehoods, or to alternative relative truths. It is not much of an accomplishment, in short, for a statement to achieve the status of relative truth; the cognitive preferability the absolutist claims for truths over falsehoods is lost for the relativist. To label *p* relatively true is not in any way to praise it or to acknowledge it to be cognitively preferable to rival propositions; it is only to register that someone believes it.[48]

What then of relativism? Meiland seeks to develop a tenable conception of relative truth, and to defend relativism as itself relatively true. But, granting for the moment the tenability of the concept, Meiland's thesis comes simply to the claim that he believes it; that relativism corresponds-to-reality-for-Meiland. Let us grant this thesis as well. Does it follow that relativism is enhanced in cognitive status, or that absolutism suffers? Not in the least. If relativism is only relatively true, then by its own lights it is no better than its alternatives. Meiland is seeking to defend relativism, but the very notion of rational defense is given up by the relativist, for the relativist has rejected the possibility of nonrelative criteria by which rival claims or hypotheses can be evaluated. Thus to hold that relativism is relatively true is not in any way to suggest that there is good reason for being a relativist, nor is it to discredit absolutism. Defending relativism as relatively true is not *defending* it at all; nor is it asserting anything about the *rightness* of relativism. Even if we grant the relativist the coherence of the concept of relative truth, then, a relativism based upon this concept does not avoid the self-refutation problem. The *UVNR* argument remains a powerful argument for the incoherence of relativism: one cannot embrace relativism as right without giving up the very notion of rightness; to defend relativism as relatively true or right is not to defend it at all. The problem here is not with the *formulation* of the concept of relative truth, but rather, with the impossibility of the *defense* of a relativism which depends upon that concept.

THE IMPOTENCE OF RELATIVISM

These considerations show, I think, the inability of relativism to sanction significant judgments, to recognize or do justice to the notions of warrant and justification with respect to beliefs and knowledge-claims, or to offer direction with respect to action. This inability may be thought of as the *impotence* of relativism. Since to say of *p* that it is relatively true is not to praise *p* or claim it to be more worthy of belief than rival claims not-*p* or *q*, and since the relativist recognizes this feature of relativism – it is true for her, since, being a relativist, she believes that truth is relative – then the relativist must realize, from her own relativistic point of view, that *p*'s being true-for-her in no way warrants belief in *p* or establishes *p* as in any way more worthy of belief than not-*p* or *q*. For the relativist must realize that not-*p* and *q* are also relatively true, and so *p* has no claim to cognitive superiority or worthiness of belief on grounds of *its* relative truth. Moreover, the relativist must realize the thrust of the *UVNR* argument, which (in this context) comes to the view that the very notion of cognitive worthiness or superiority is given up with the adoption of relative truth.

Consequently, the relativist cannot say "I should adopt the belief *p*," or "*p* is justified for me, because *p* is true for me," because *p*'s relative truth in no way – from the relativist's own point of view – sanctions *p* or affords *p* epistemic warrant of any sort. Thus the relativist cannot regard her beliefs, or her relative truths, as warranted or worthy of belief. Similarly for action. The relativist is thus left in the position of being unable to recognize relative merits of propositions, knowledge-claims, or actions – all are as worthy as the rest, and the very notion of worthiness has been jettisoned – and therefore cannot rationally prefer any relative truth to any other. Warranted adoption of belief and of courses of action is impossible for the relativist. Here is relativism's impotence.

Of course the relativist can respond by claiming that *p*'s being true-for-*her* makes it, though only relatively true, more worthy of belief for her than a rival proposition *q*, which, while relatively true, is true-for-*someone-else*. But unless the relativist can say why a relative truth which is true-for-her is superior to or more worthy of belief than a relative truth which is true-for-someone-else – which she can't, since, in embracing relativism, she has given up the very notion of worthiness of belief – then adoption of one relative truth over another on the grounds that the adopted truth is true for the person doing the adopting is arbitrary. *p*'s truth-for-Jones is not in any way superior to not-*p*'s truth-for-Smith – even for Jones. Thus the relativist can adopt beliefs (and act) only arbitrarily. Either arbitrariness or impotence – these are the sole options of the relativist.[49]

RELATIVISM AND ARGUMENT

Another response Meiland can make to the *UVNR* argument is to claim that, in 'defending' relativism and the coherence of the concept of relative truth, he is not *defending* relativism or *arguing* for it at all, but rather is attempting simply to articulate his view, present it in a logically ordered way, or meet some purpose other than that of rational persuasion. Meiland actually does offer this response.

Maurice Mandelbaum argues that epistemological relativism is 'self-limiting'; i.e. that "acceptance of relativism in the theory of knowledge frequently – and perhaps always – involves a prior commitment to non-relativistic interpretations of at least *some* judgments concerning matters of fact."[50] Mandelbaum's argument centers on the point that, in order for the relativist to defend or argue for relativism, she must regard her evidence non-relativistically. If she does not, Mandelbaum suggests, her evidence need not be regarded *as evidence* by the nonrelativist; consequently, the evidence will not carry the cognitive force of evidence, and so will not effectively persuade the nonrelativist of the rightness of relativism.

In considering Mandelbaum's argument, Meiland writes:

> Mandelbaum's basic objection is that the relativist offers arguments to support his position and, in doing so, presupposes some statements to be absolutely true . . . throughout this discussion so far, an assumption has been made – namely that when the relativist gives arguments to the nonrelativist, he is trying to convince the nonrelativist. Persuasion may be the nonrelativist's purpose in giving arguments, but it need not be the relativist's purpose. The giving of arguments can have many purposes. For example, the relativist

may use the argumentative form in order to present his view in a logically ordered way so that his audience will better understand that view.

To insist that the relativist should have the same purposes in argument as the nonrelativist might have is misguided. It is an example of a practice often followed by absolutist objectors to relativism – the practice of requiring the relativist to adopt ends and to satisfy standards which are appropriate to absolutism and then declaring relativism refuted when relativism fails to live up to the mark. Relativism may be more appropriately considered as a world-view which generates its own goals and standards.[51]

Meiland does not here actually deny that his many arguments offered in support of relativism and in opposition to absolutism are meant to provide reasons for adopting the former and rejecting the latter, but the implication of the passage is clear. Meiland is suggesting that he is not trying to provide non-relative reasons in support of relativism, or to defend relativism as right or as cognitively superior to absolutism. In this way Meiland seems to meet Mandelbaum's argument – the relativist is not contradicting herself in adducing 'evidence' for relativism, for her arguments and her evidence are not to be regarded as efforts at rational justification of relativism which are intended to be cognitively forceful for the absolutist as for the relativist. Similarly, Meiland seems to be speaking to the *UVNR* objection as well, since he seems to be claiming that he is not engaged in the project of establishing the *rightness* of relativism.

Unfortunately, this approach will not rescue Meiland from the difficulties delineated thus far. For one thing, it belies Meiland's own arguments, which he seemingly offers (contrary to the suggestion now being considered) not for the purpose of articulation or some other purpose, but for the purpose of convincing or rationally persuading the reader of the rightness of his views. When Meiland writes, for example, that "the relativist's position is surprisingly strong and is not touched by the usual accusations of incoherence, infinite regress, and begging of the question,"[52] or when he summarizes an article by writing that he has tried "to *show* that the relativist has answers to [critical] points . . . [and he has] tried to *show* that the charge of self-vitiation against relativism is much harder to substantiate than the critics seem to have thought,"[53] or when he summarizes another article by claiming that the discussion therein sufficies to *refute* the charge of self-refutation,[54] it is difficult to interpret these remarks, and Meiland's argumentative efforts, in any other way than as remarks and efforts aimed at rationally establishing and persuading the reader of the cogency of relativism and the failure of criticisms of relativism. He is engaged in the project of providing evidence for, and reason for adopting, relativism (or at least not rejecting it for what he thinks he has shown are spurious reasons). Thus despite his remarks about relativistic purposes of argument other than that of establishing evidence and persuasive cases, his own arguments undoubtedly are intended to satisfy these 'absolutist' purposes of argument. Consequently, Meiland's arguments *do* open themselves up to the challenges posed by Mandelbaum's self-limiting argument and the *UVNR* argument. Moreover, it is difficult to know what argument comes to if it does not involve the effort to establish a conclusion on the basis of reasons – which is to say, to make a persuasive case for the rightness of some conclusion. It is not clear why discussions without this cognitive aim should be regarded as *arguments* at all. The relativist who is not seeking with her arguments to establish the rightness of relativism simply is not engaging the issue. So Meiland's suggestion that the relativist should not be seen as engaged in the project of defending relativism or advancing arguments with the purpose of establishing reasons for embracing or making rationally compelling cases for relativism fails to rescue the relativist either from Mandelbaum's self-limiting argument or from the *UVNR* argument.

DID SOCRATES BEG THE QUESTION?

If the points made thus far are correct, the relativist has failed to secure for the concept of relative truth a modicum of intelligibility, and has failed to meet the challenge of the arguments for the incoherence of relativism. Because these arguments trace their lineage to Socrates' discussion of Protagoras in the *Theaetetus*, it is important to consider the charge, put forward by several defenders of relativism, that Socrates' arguments beg the question against the relativist and so fail to demonstrate the untenability of relativism. We must consider both whether Socrates' arguments so beg the question, and also whether more recent formulations of the arguments for self-refutation do so.

It has long been recognized in the literature that Socrates' arguments seem to beg the question, by dropping the relativizing phrase 'for . . .' when discussing truth, thus collapsing the Protagorean notion of relative truth into the standard absolutist one. Thus Meiland writes that

> Plato's own attempt, in the *Theaetetus* to show Protagorean relativism to be self-refuting appears to be radically defective due to Plato's dropping of the relativistic qualifier (the "for me" in "true for me") at crucial points.[55]

Chris Swoyer similarly concludes that

> such criticisms beg the question against relativism in their implicit reliance upon an absolute theory of truth.[56]

James N. Jordan also argues that, in response to Socrates' arguments,

> Protagoras would doubtless reply, and rightly, that this leaves him untouched, that in fact it begs the question. . . . "For . . ." is an addendum whose power Socrates has misjudged, if, indeed, he has not entirely overlooked it.[57]

M.F. Burnyeat similarly recognizes a problem for Socrates here.[58]

It is undeniable that Socrates does drop the qualifier 'for . . .' in several passages in his response to Protagoras. The crucial question is, to what extent does this deletion vitiate his criticisms of Protagorean relativism? The answer, I think, is that the deletions do not in the end limit the force of his critique, or of the present reconstruction of it.

First, it must be noted that the Socratic argument can itself be analyzed in such a way that a portion of the argument includes a demonstration of Protagoras' commitment to and reliance upon the notion of absolute truth, and consequently of the incoherence of his concept of relative truth.[59] If Socrates' argument shows this, then it does not beg the question against Protagoras by deleting the qualifier 'for . . .', but rather deletes it for the very good reason that the expression 'true for' is only intelligible when taken to be equivalent to 'true (simpliciter)'.[60]

But whether the arguments in the *Theaetetus* can be shown to be compelling, the reconstructions of them offered here clearly can. The *UVNR* argument has not been satisfactorily fended off, either by Protagoras or by the contemporary relativists considered thus far; similarly, the *NSBF* argument stands as a powerful refutation of relativism. For, as we have seen above, the relativist cannot coherently assert relativism only relatively: to argue that relativism is only correct for the relativist is to fail to join the issue with the opponent of relativism; it is to fail to assert the correctness or cognitive superiority of relativism. But to defend or assert relativism non-relativistically is to acknowledge the cognitive force of criteria or principles of reasoning by which the relativistic thesis can itself be assessed, and this acknowledgement constitutes a rejection of the relativistic thesis (which rejects the force of such criteria or principles) which is purportedly being defended. Thus the relativist cannot coherently defend or assert relativism either relatively or non-relatively. And, as the *NSBF* argument illustrates this dilemma for the relativist, if the relativist defends relativism relativistically, she recognizes the equal cognitive legitimacy of absolutism (and the standards by which absolutism is (relatively) established as superior to relativism) and thus the non-superiority of relativism and the arbitrariness of her commitment to it; while if she defends relativism non-relativistically she gives up the very doctrine she is attempting to defend.[61] Both the *UVNR* and *NSBF* arguments proceed without a commitment to absolute truth, but rather portend trouble for the relativist whatever conception of truth she adopts, and so these arguments do not beg the question against the relativist.[62]

The final telling point against the charge that the absolutist (Socrates) begs the question against the relativist (Protagoras) by assuming a notion of absolute truth (deleting the qualifier 'for . . .') is that the qualified, relativistic concept of truth has yet to be made sense of. Neither Protagoras nor the contemporary relativists considered above have offered an analysis of the concept of relative truth (or the qualifier 'for . . .') which renders that concept intelligible, except by trivializing it by reducing it to 'mere belief' (so that '*p* is true for X' just comes to 'X believes that *p*') or rejecting it by demonstrating its reliance on the concept of absolute truth (so that '*p* is true for X' comes to '*p* is (absolutely) true for X'). But if the concept of relative truth is either incoherent, trivial, or reliant on the concept of absolute truth, it is hardly a powerful criticism of the absolutist (Socrates) that she fails to acknowledge the legitimacy of the concept of relative truth and utilizes an absolutist concept (drops the qualifier 'for . . .') in analyzing the relativist's position. In this circumstance, the absolutist can hardly be seen as begging the question against the relativist. On the contrary, she must be seen as doing her opponent a favor. The force of the question-begging charge, in short, depends on the coherence of the concept of relative truth; and we have been given no reason to regard that concept as coherent – except in the special cases in which it is trivial or parasitic upon the absolutist conception. For these reasons, it seems appropriate to conclude that the question-begging charge does not constitute a major difficulty for the absolutist.

RELATIVISM WITHOUT RELATIVE TRUTH

Up to this point we have been concerned mainly with the difficulties attending versions of epistemological relativism which incorporate or rely upon the concept of relative truth. A very different sort of relativism is that which eschews that concept. In this section I want briefly to consider a recent defense of relativism which explicitly rejects relative truth, that of Hartry Field.

Field's discussion of relativism takes place in the context of his critique of Hilary Putnam's recent work.[63] As is well known, Putnam has recently given up his earlier embrace of 'metaphysical' realism, and has embraced what he calls 'internal' realism. In the course of his discussion of realism, Putnam forcefully rejects epistemological relativism as incoherent.[64] Field seeks to defend metaphysical realism from Putnam's critique; in addition he offers a conception of epistemological relativism which he believes shows relativism to be "a coherent and palatable doctrine."[65] The relativism Field defends "is a relativism as to values only, not to facts."[66] Being a physicalist and a metaphysical realist, Field regards truth as factual and thus non-relative; there is therefore no sanction of relative truth in Field's relativism. There is, however, relativism about values, including especially *epistemic* values: "to say that a belief is justified is to evaluate it."[67] Field articulates his view further as follows:

> I understand it [epistemological relativism] as the doctrine that the basic epistemological properties are not such properties as that of belief *B being justified*, but rather such relativized properties as that of belief *B being justified relative to* evidential system *E*. (An *evidential system* is, roughly, a bunch of rules for determining under what conditions one is to believe various things; a belief is justified relative to an evidential system in certain circumstances if the rules license the belief under those circumstances.)[68]

Field is quick to point out that, while relativism so conceived allows the relativist to say that *B* is justified relative to E_1, or not-*B* relative to E_2, she cannot say that *B* is justified *simpliciter*, or to *the true* evidential system, for these notions of justification are unrelativized and so are not available to the relativist. Field's relativist can say that *B* is true 'absolutely', but that *B* is justified only relatively to some *E*.

The obvious question to be put to Field's brand of relativism is: can rival, incompatible evidential systems themselves be non-relatively or objectively or rationally evaluated? For if they can, so that we can say that E_1 is a better or superior or more justified evidential system than E_2 absolutely or non-relatively, then the relativism Field offers is not very relativistic – for while two relativists who hold different *E*'s can claim that their incompatible beliefs *B* and non-*B* are justified relative to their respective *E*'s, we can nevertheless judge (say) that *B* should not be believed since it is justified only relative to E_1, which we can argue is inferior to, or less adequate than, E_2. This would reduce Field's relativism to a species of absolutism, and seems in any case to be incompatible with Field's view that epistemic values are relative – for in judging *E*'s we are, after all, *evaluating* *E*'s, and such evaluating, no less than evaluating beliefs, is epistemic evaluating and so should be relativized (to 'meta-*E*'s') on Field's view. On the other hand, if rival, incompatible *E*'s cannot be non-relatively evaluated, then it is difficult to see how any belief, no matter how bizarre, can be ruled out or evaluated negatively, for some *E* which will sanction it could always be constructed. Consequently, the relative evaluation of beliefs will be empty – *all* beliefs will be justified relative to some *E*'s, unjustified relative to others, and the notion of epistemic justification will be not simply relativized, but trivialized; to say that a belief is justified will be not to praise it at all. The impotence noted earlier with respect to relative truth resurfaces in full force with respect to relative epistemic evaluation.

Field might bite the bullet at this point and accept the impotence attending his thesis of relative epistemic evaluation. But in fact he takes the opposite course, and, in a crucial passage which draws upon his distinction between relative evaluation and non-relative facts, argues that on his conception the relativist is not so impotent with respect to epistemic judgment as she appears to be:

> The idea that a relativist can do nothing more than say things like "*B* is justified relative to my evidential system but not to yours" is the basis for one of Putnam's main complaints against relativism. (Cf. 119/20.) In fact, however, an epistemological relativist is not nearly so limited. For evidential systems have lots of factual properties besides the properties of being adequate idealizations of the epistemic behavior of particular people, and the relativist can use such further factual properties of evidential systems in his epistemological remarks. For instance, one factual property that an evidential system may have is *reliability*, i.e., the property of tending to lead to a high proportion of truth over falsity in the long run. This talk of reliability is vague; but, vagueness aside, it is talk that the epistemological relativist is free to indulge in, for it does not involve unrelativized epistemological notions. In particular, an epistemological relativist can want, and presumably will want, to bring his belief-forming behavior into accordance with some reliable evidential system. Consequently the highest epistemological praise will not be "is justified relative to my evidential system"; if there is any "highest epistemological praise" it will be something like "is justified relative to some highly reliable evidential system" (or "is justified relative to all highly reliable evidential systems," or some such thing). This isn't really an adequate formulation of what "the highest epistemological praise" (if there is such a thing) would be, for (among other things) reliability is not the only feature we want our evidential systems to have; but it gives the general flavor.
>
> The above remarks should make clear that (contra Putnam) an epistemological relativist will not hold that all evidential systems are equally good. . . .[69]

Grant Field that epistemic systems have factual properties, and that on his conception of relativism the relativist is free to appeal to such properties. Does it follow that the relativist "will not hold that all evidential systems are equally good"? No. Consider Field's example: as a matter of fact, some *E*'s will be more reliable than others. Suppose that E_1 and E_2 are similar with respect to most factual properties; they differ only in that E_1 is more reliable than E_2. Must the relativist regard E_1 as better than E_2? No. For, whatever the factual properties of E_1 and E_2, the relativist is free to *evaluate* those factual properties in whatever fashion she wishes (or, strictly speaking, in accordance with her evidential 'meta-system' for evaluating evidential systems). Reliability may be a factual property of an evidential system, but *valuing* reliability is not factual, but evaluative;[70] and by Field's own lights the relativist's evaluations are relative to her evidential system (or, again, her meta-system for evaluating first-order systems). Thus there is no reason to suppose that the relativist would value a reliable system more than, say, a system which was unreliable, or which regularly produced a good laugh. When Field says that the relativist "presumably *will* want to bring his belief-forming behavior into accordance with some reliable evidential system,"[71] he is presuming that the relativist values reliability – but such a presumption is precisely what is ruled out by the view that epistemological evaluations are relative. Some (meta-)systems will value reliability, others will not; and a consistent epistemological relativism with respect to values must not prejudge the evaluations of alternative (meta-)systems; nor can it evaluate those evaluations. Judgments of the goodness of evidential systems are epistemological evaluations, and these are relative to evidential (meta-)systems; therefore Field's relativist, who holds that epistemological evaluations are relative, must hold that judgments of evidential systems, which are epistemological evaluations, must likewise be relative, and that rival evidential meta-systems will evaluate rival evidential systems differently (and that these evaluations are themselves relative to 'meta-meta-systems' and so cannot be judged 'absolutely', etc.). Consequently, the epistemological relativist *will* hold, indeed *must* hold, contra Field, that "all evidential systems are equally good." On Field's position there is no room for *any* sort of 'epistemological praise', for the very notion of epistemic worth is trivialized (by being relativized). The evaluation of evidential systems is just as much a matter of epistemological evaluation as the evaluation of beliefs; the relativism Field extends to the latter must be extended to the former as well. But to evaluate systems as Field does in the cited passage (e.g. with respect to reliability) is to evaluate systems non-relatively – and this is just what the relativist cannot do. Thus Field has not defused Putnam's complaint, which is tantamount to the charge of impotence noted earlier. Field's version of relativism is open to the *UVNR* argument as well, since there are no restraints on the valuations of rightness of alternative incompatible evidential systems (or meta-systems). Consequently, relativism without relative truth seems to be just as untenable as the more typical sort of relativism which incorporates some conception of relative truth. So long as evidential systems cannot themselves be neutrally or non-relatively evaluated,[72] the incoherence problems attending relative-truth versions of relativism apply to Field's physicalist, absolute-truth version as well. On the other hand, were Field to hold that evidential systems are capable of non-relative evaluation,[73] he would be in effect giving up relativism altogether. In this respect as well Field's absolute-truth version of relativism fares no better than its relative-truth cousins.

Field has suggested[74] that I have blurred a crucial distinction which, if respected, would overcome the criticism of his version of relativism just adumbrated. This distinction is between (relative) *evaluation* and (non-relative) *preference*. While epistemological judgments (concerning e.g. the justification of beliefs) are relative, on Field's view, to evidential systems, our judgments concerning evidential systems themselves are not evaluations, but rather are preferences which reflect our epistemic goals. If my epistemic goals include (say) reliability, and evidential system *E* has the property of reliability, then I may prefer *E* to alternative systems which do not have that property (or have it to a lesser degree).[75] In this case my epistemological judgments are relative to *E*, but my adopting *E* is not a matter of evaluation of rival systems, but rather a reflection of my preference, which is itself a function of my epistemic goals. (Preferring *E* involves (at least in part) believing that *E* best achieves one's epistemic goals. This ability to achieve epistemic goals is objective and factual, so one might be wrong about (say) *E*'s being more reliable than *E*′, and one might disagree with one's opponent regarding evidential systems despite agreement about goals. But often disputes about evidential systems reflect disputes about epistemic goals.)

I do not see that the distinction between evaluation and preference helps Field's relativism. Consider the case of rival, incompatible epistemic goals. On Field's view, one's preference for an evidential system with a high degree of reliability is not an evaluation which is relative to one's evidential (meta-)system; it is simply an expression of one's epistemic goals. Suppose Smith's and Jones' epistemic goals differ, so that Smith adopts evidential system E, which (she thinks) best reflects her non-relative preferences, while Jones adopts E', which (she thinks) best reflects *her* preferences. Can we evaluate or assess these preferences? If we can, then those preferences are subject to evaluation by means of some evidential system, and so on Field's view are relative. If we cannot, then the preferences must be seen as unsupported, unjustified, and unfounded – you have your preferences, I have mine – and a relativism of the most crass sort ensues. Preferences on this view are objective, in the sense that it is either true or false that Smith has preference P; but they are surely relative nonetheless – not because they are epistemological evaluations which are relative to one's evidential system, but because they are relative to one's personal predilections and are in no way supported or defended as appropriate or reasonable (let alone correct).

The crucial question is: (how) can one evaluate preferences for evidential systems which reflect epistemic goals? Field's distinction between evaluation and preference does not show (how or) that one can evaluate preferences, in which case epistemological judgments are relative to systems which reflect unsupported preferences and goals. And this is just the sort of relativism criticized in the text.

Lest one suppose that it is somehow meaningless or inappropriate to evaluate preferences or epistemic goals, I hasten to point out that it is not only meaningful to do so, but that it is often of prime epistemological concern. Consider, for example, Ronald, a follower of Ptolemy who believes that the earth is at the center of the solar system, which belief is licensed by his evidential system E. E is embraced by Ronald because it reflects his epistemic goals, which include those of preserving the data provided by ordinary (naive realist) perception (e.g. that the sun moves across the sky; that the earth does not move; etc.), and of maintaining a worldview which places humans at the center of God's universe. Confronted with Ronald, we have two choices: say that his beliefs are as (relatively) justified as our own, because they are licensed by his evidential system, which reflects his epistemic goals and preferences; or else challenge those beliefs by challenging the E and the set of goals and preferences which license those beliefs. If we opt for the former, we embrace relativism; if we opt for the latter, we hold that epistemic goals and preferences are themselves capable of epistemological evaluation and assessment. Either way Field's distinction between evaluation and preference does not save his position. In the former case, preferences are not evaluated with reference to evidential systems, and relativism reigns but is open to the criticisms developed above. In the latter case preferences and epistemic goals are themselves open to evaluation and so are not relevantly distinguishable from evaluations.

CONCLUSION

I conclude that the defenses of relativism considered thus far have been successfully undercut. More specifically, the arguments for the incoherence of relativism are as compelling as ever, and have manifestly not been laid to rest by contemporary relativists. The basic Socratic insight that relativism is self-refuting, and so incoherent, remains a fundamental difficulty for those who would resuscitate and defend the ancient Protagorean doctrine or a modern variant of it.

But, of course, there are many other avenues to relativism yet to be considered. In the next chapter I turn our attention to relativistic arguments which hinge on the notions of conceptual scheme or framework.

CHAPTER 2

FRAMEWORKS, CONCEPTUAL SCHEMES, AND "FRAMEWORK RELATIVISM"

INTRODUCTION

Chapter 1 was concerned primarily with the question of relativism's alleged self-refutation. The present chapter takes as its subject the cluster of arguments for and against relativism which focus on the idea that cognitive activity is inevitably conducted from some ongoing perspective or point of view.

In the literature on relativism, considerable attention has been paid to the kindred notions of "conceptual framework" and "conceptual scheme." The relationship between these notions and the doctrine of relativism is, nonetheless, problematic. Popper, for example, regards the notion of conceptual framework or scheme as coherent, but the associated epistemological doctrine as false. Davidson, as is well known, regards the notion, and the attendant scheme/content distinction, as unintelligible, and epistemological relativism as an incoherent doctrine. Rorty accepts Davidson's argument against the intelligibility of the notion, then proceeds to build upon that argument an epistemological position as relativistic as one could want. How is one to make sense of this morass of incompatible analyses?

In this chapter I try to sort out the several claims made on behalf of the (in)coherence of the notion and of the (in)coherence or truth (falsity) of the doctrine. In particular, I will be concerned with the question to what extent (if any) the notion, if coherent, lends support to epistemological relativism. I will argue that the notion is coherent, contra Davidson, but that the doctrine of epistemological relativism gains no support from that coherence. I begin by articulating the notion with a bit more precision, so that we might have a better idea of what notion it is whose coherence is being examined.

FRAMEWORKS, SCHEMES, AND THE BOUNDEDNESS OF JUDGMENT

There are a host of alternative renditions of the notion (which I will refer to both by "framework" and "scheme," so as to reflect the favored usage of the several authors to be considered) in the literature.

Weinert[1], for example, notes that a variety of forms of relativism all involve the claim that

> the standards available to assess the value of the activity under review are, in some sense, internal to a given framework of reference which cannot, by the act of assessment, be transcended. Thus relativism implies the existence of some kind of framework – be it social, epistemological or linguistic – within which activities or entities can (comfortably) be judged rational, moral or beautiful but which, at the same time, constitute boundaries beyond which the assessments cannot extend without clashing with standards derived from different frameworks.

Popper[2] defines relativism as the view that

> truth is relative to our intellectual background or framework: that it may change from one framework to another.

Meiland and Krausz[3] write that

> In one of its most common forms cognitive relativism holds that truth and knowledge are relative, not to individual persons or even whole societies, but instead to factors variously called conceptual schemes, conceptual frameworks, linguistic frameworks, forms of life, modes of discourse, systems of thought, Weltanschauungen, disciplinary matrices, paradigms, constellations of absolute presuppositions, points of view, perspectives, or worlds. What counts as truth and knowledge is thought to depend on which conceptual scheme or point of view is being employed rather than being determinable in a way which transcends all schemes or points of view.

Such quotations could be multiplied indefinitely, but there would be no point in doing so. What the several citations make clear[4] is that central to the standard defense of what may be called "framework relativism" is the notion that epistemic judgments are in some sense *bound* by schemes or frameworks, so that cognizers are limited or trapped by, and cannot transcend or escape from, some sort of fundamental restraints which sharply delimit the possible range of claims they are able to regard as true or justified. It is the idea that there is a boundary beyond which defensible judgments cannot be made, and an arbitrariness or in-principle-unjustifiedness about the particular features of any given boundary, that is the heart of framework relativism; and "conceptual scheme" and "conceptual framework" are phrases intended by relativists to point to the existence, and specific features, of such boundaries or limit conditions. Framework relativism is thus dependent on the notion of a limit or boundary beyond which rationally defensible judgments concerning truth or epistemic worthiness cannot be made. One might judge in terms of the criteria operative in one's own framework or scheme, but one cannot non-question-beggingly judge those criteria, or one's own scheme or framework, to be epistemologically superior to alternative criteria, frameworks or schemes. For such judgment would require the transcendence of one's own scheme in order neutrally to compare it with alternatives – and this is, according to the framework relativist, precisely what cannot be done. The limit, or boundary, of one's framework or scheme is located just at the point of scheme transcendence: one can perfectly well judge from within one's scheme, utilizing criteria internal to the scheme, but one cannot meaningfully question the scheme or its criteria themselves, for they are necessary for judgments to be made at all. The force of the idea that we are trapped within our conceptual schemes or frameworks is precisely that we cannot escape, transcend, or get outside them in order fairly, neutrally, or non-question-beggingly to assess them. Such a possibility might well be called by framework relativists "the myth of absolutism." There simply is not, according to the relativist, any framework- or scheme-independent vantage point from which to criticize or judge alternative frameworks or schemes.[5]

Below I will consider Davidson's claim that the very notion of conceptual scheme is unintelligible. First, however, I would like to turn our attention to Popper, who grants the intelligibility of the notion, but argues that the notion is in fact inapplicable to our cognitive situation. That is, Popper claims that, while we do operate from within frameworks, it is simply false that we are trapped by them or incapable of transcending, criticizing, or neutrally judging them.

POPPER AND PICKWICKIANISM

According to Popper, while it is true that our cognitive activities are in some sense functions of our conceptual frameworks, we are nevertheless perfectly capable of studying those frameworks and the influence they have on our views, of objectively criticizing them, and of rationally opting for changes, which may rightly be regarded as improvements, in them. Consequently, it is simply false that we are trapped by our frameworks:

> . . . the relativistic thesis that the framework *cannot* be critically discussed is a thesis which *can* be critically discussed and which does not stand up to criticism.
>
> I have dubbed this thesis *The Myth of the Framework* . . .

> I do admit that at any moment we are prisoners caught in the framework of our theories; our expectations; our past experiences; our language. But we are prisoners in a Pickwickian sense: if we try, we can break out of our frameworks at any time. Admittedly, we shall find ourselves again in a framework, but it will be a better and roomier one; and we can at any moment break out of it again.
>
> The central point is that a critical discussion and a comparison of the various frameworks is always possible. . . .
>
> The Myth of the Framework is, in our time, the central bulwark of irrationalism. My counter-thesis is that it simply exaggerates a difficulty into an impossibility.[6]

As these passages make clear, Popper sees no problem with the intelligibility of the notion of conceptual framework. Indeed, he thinks it true that we are "prisoners caught in the framework of our theories; our expectations; our past experiences; our language." But he denies that our imprisonment fosters epistemological relativism, for we can always escape from the prison of our framework. We can get outside of our framework, and question, examine, criticize, and improve it. Of course getting outside of our present framework is not the same as getting outside of *all* frameworks, and Popper does not claim that we can arrive at a totally framework-transcendent perspective: to escape our own is to land us in another. But this other will be, or at least can be, better ("roomier"); and we can in turn escape from, and criticize, this newer framework. In this way Popper affirms that objective, non-question-begging evaluation of frameworks is possible, and denies that we are trapped in our frameworks in the way that the "myth of the framework," and the "framework relativist" (i.e. the epistemological relativist who bases her relativism upon considerations of frameworks and schemes), suggest.

Popper grants the facts of cultural diversity which lead some to formulate framework relativism. He quotes from Xenophanes the following 2500 year old articulation of these facts[7]:

> The Ethiops say that their gods are flat-nosed and black,
> While the Thracians say that theirs have blue eyes and red hair.
> Yet if cattle or horses or lions had hands and could draw,
> And could sculpture like men, then the horses would draw their gods
> Like horses, and cattle like cattle; and each they would shape
> Bodies of gods in the likeness, each kind, of their own.

While Popper grants the facts of diversity and bias that Xenophanes here points to, he emphatically denies the relativistic conclusion which some have drawn from them[8]:

So each of us sees his gods, and his world, from his own point of view, according to his tradition and his upbringing; and none of us is exempt from this subjective bias.

This argument has been developed in various ways; and it has been argued that our race, or our nationality, or our historical background, or our historical period, or our class interest, or our social habitat, or our language, or our personal background knowledge, is an insurmountable, or an almost insurmountable, barrier to objectivity.

The facts on which this argument is based must be admitted; and indeed, we can never rid ourselves of bias. There is, however, no need to accept the argument itself, or its relativistic conclusions. For first of all, we can, in stages, get rid of some of this bias, by means of critical thinking and especially of listening to criticism. For example, Xenophanes doubtless was helped, by his own discovery, to see things in a less biased way. Secondly, it is a fact that people with the most divergent cultural backgrounds can enter into fruitful discussion, provided they are interested in getting nearer to the truth, and are ready to listen to each other, and to learn from each other. This shows that, though there are cultural and linguistic barriers, they are not insurmountable.

Thus it is of the utmost importance to profit from Xenophanes' discovery in every field; to give up cocksureness, and become open to criticism. Yet it is also of the greatest importance not to mistake this discovery, this step toward criticism, for a step towards relativism. If two parties disagree, this may mean that one is wrong, or the other, or both; this is the view of the criticist. It does not mean, as the relativist will have it, that both may be equally right.

FOR WANT OF AN ARGUMENT

I applaud the sentiment Popper here expresses. But it must be admitted that the argument Popper offers is weak and uncompelling. The framework relativist will surely not grant that we can shed our biases in stages by listening to criticism – for she will claim that what we view as a shedding of biases is itself a judgment biased by our framework. Moreover, she will object to the very notion of bias as Popper uses it, for that notion assumes, question-beggingly, that a framework-neutral unbiased vantage point is possible, a possibility the framework relativist denies. Similarly, Popper's claim regarding the factuality of fruitful discussion depends on question-begging non-relativistic conceptions of fruitfulness and of truth, both of which the relativist will reject. Thus, while Popper's discussion articulates well the denial of framework relativism, it does not constitute any sort of powerful argument for his non-relativistic stance. Indeed, Popper's discussions of the myth of the framework and of relativism are uncharacteristically non-argumentative. In his most systematic discussion, he motivates his consideration of relativism with the claim that[9]

One of the more disturbing features of intellectual life at the present time is the way in which irrationalism is so widely advocated, and irrationalist doctrines taken for granted. In my view, one of the main components of modern irrationalism is relativism.

Popper continues on to say that he will "explain *and criticize*"[10] the myth of the framework, upon which (he thinks) relativism rests. But his discussion is virtually entirely one of explanation – of his own views regarding criticism, rationality, truth, and the origins of criticism with the Ionian school; and, more relevant for present purposes, of the historical roots of the myth of the framework, which root in early observations of cultural differences, find a home in Hegel's philosophy, and illicitly gain in plausibility through a confusion between relativism and fallibilism and through flawed embracings of Whorfian linguistic relativity, Quinian ontological relativity, Kuhnian philosophy of science, and recent sociology of knowledge. It is only on the penultimate page of the essay that Popper terminates his explanation of the origins and popularity of the myth of the framework and offers a "logical diagnosis of the malaise."[11] Unfortunately, the diagnosis offered hardly constitutes a compelling argument against the myth. It consists in the claim that framework relativists are actually "justificationists," in Popper's sense of the term: that is, they assert that hypotheses (or theses or claims) must be proved or justified by deduction from unassailable principles or axioms; they note that justification in this sense is impossible, for alternative frameworks offer alternative, incompatible principles or axioms, and there is no getting outside of frameworks in order to critically assess alternative principles; they conclude that framework relativism (or dogmatism, or infinite regress) appears inevitable. Popper rejects this chain of reasoning by rejecting "justificationism" in favor of his favored method of criticism, conjecture and refutation. Leaving aside the devastating problems with Popper's discussion of "justificationism," this rejection hardly defeats the relativist. Popper writes that[12]

the *correct method* of critical discussion starts from the question: what are the *consequences* of our thesis or our theory? Are they all acceptable to us?

Thus it consists in comparing the consequences of different theories (or, if you like, of different frameworks) and tries to find out which of the competing theories or frameworks has consequences that seem preferable to us.

But the framework relativist will surely respond that our preferences are themselves functions of our frameworks, so that the "correct method" offers no escape from the myth. Thus Popper's hasty and superficial argument against the myth hardly constitutes any sort of compelling refutation of framework relativism.

Popper affirms the possibility of framework criticism, and so the denial of framework relativism, in many places. But he offers no good argument for his view. In fact, his discussion in "The Myth of the Framework" consists almost entirely of an explanation of the origins and pervasiveness of the myth; the poor and hasty argument he throws in on the last two pages seems to be little more than an afterthought[13] and at any rate fails utterly to support his rejection of the myth and of framework relativism. Popper's other discussions of the myth and of relativism[14] do no better. So an argument against the myth, and framework relativism, is still wanted. Davidson's view that framework relativism (or, since he uses "scheme" rather than "framework," "scheme relativism") is not false, but incoherent, is often taken as such an argument.

DAVIDSON AND THE SCHEME/CONTENT DISTINCTION

Popper and Davidson both reject scheme relativism. But their rejections are very different. For while Popper regards the doctrine as false, Davidson regards it as unintelligible, for it rests upon an unintelligible distinction between conceptual schemes and some sort of scheme-neutral content.

Davidson's conception of conceptual schemes is akin to the ones we have been considering, as is the way in which it is alleged to give rise to relativism[15]:

> Conceptual schemes . . . are ways of organizing experience; they are systems of categories that give form to the data of sensation; they are points of view from which individuals, cultures, or periods survey the passing scene. There may be no translating from one scheme to another, in which case the beliefs, desires, hopes and bits of knowledge that characterize one person have no true counterparts for the subscriber to another scheme. Reality itself is relative to a scheme: what counts as real in one system may not in another.

But Davidson denies that this conception of conceptual schemes is intelligible (and, consequently, that scheme relativism is a coherent doctrine). For it relies on the possibility of alternative schemes, untranslatable into one another and so incomparable with one another. Equating schemes with language, and focusing on the relation of translatability, Davidson reformulates the question "Can alternative conceptual schemes fail of comparability?" as "Can alternative languages fail of translatability?". He argues that they cannot. Considering first the possibility of complete failure of translatability, in which "no significant range of sentences in one language could be translated into the other"[16], Davidson argues that this possibility rests on the untenable dualism of conceptual scheme and unconceptualized content: difference of conceptual scheme rests on failure of intertranslatability; unconceptualized content, which is neutral to and outside of schemes, is necessary for such failure:[17]

The failure of intertranslatability is a necessary condition for difference of conceptual schemes: the common relation to experience or the evidence is what is supposed to help us make sense of the claim that it is languages or schemes that are under consideration when translation fails. It is essential to this idea that there be something neutral and common that lies outside all schemes.

Thus – since a dualism of scheme and content is tantamount to the notion of unconceptualized content, such content is necessary for a failure of intertranslatability, and such failure is necessary for difference of conceptual scheme – a rejection of the initial dualism undercuts the possibility of alternative conceptual schemes. Noting that writers such as Whorf, Kuhn, Feyerabend, and Quine all rely on the scheme/content distinction, on the idea that conceptual schemes (languages) stand in some particular relation to experience, Davidson argues that this idea, and the attendant distinction, should be rejected. For making sense of the idea (and the distinction) requires acceptance either of the analytic/synthetic distinction and a "fixed stock of meanings," or of a divorce between the notions of truth and translation. Assuming the untenability of the first, and arguing for the untenability of the second,[18] Davidson concludes that the scheme/content distinction must be rejected – and with it, any hope of rescuing the notion of alternative conceptual schemes[19]:

Neither a fixed stock of meanings, nor a theory-neutral reality, can provide, then, a ground for comparison of conceptual schemes. It would be a mistake to look further for such a ground if by that we mean something common to incommensurable schemes. In abandoning this search, we abandon the attempt to make sense of the metaphor of a single space within which each scheme has a position and provides a point of view.

There is then, according to Davidson, "no intelligible basis on which it can be said that schemes are different."[20] No scheme-neutral content or scheme/content distinction, thus no failure of intertranslatability, thus no difference of conceptual scheme, thus no coherent conception of scheme relativism – this is the structure of Davidson's argument. There cannot be, on this view, alternative conceptual schemes, since to be a genuine scheme (language) a putative scheme must be translatable into our language – in which case it is not an *alternative* to our own – and by the same token, if a putative scheme fails of intertranslatability, we have no reason to regard it as a scheme at all.[21] And, to keep the issue of scheme relativism before us, I remind the reader that if the notion of alternative conceptual schemes is unintelligible, as Davidson's argument seems[22] to show, than scheme relativism is lost as well, since such relativism requires that something – truth, knowledge, reality, or some such – is relative to alternative schemes: if no alternative schemes, then no scheme relativism. Thus Davidson's argument, though very different in substance from Popper's, is no more welcomed by the friend of relativism.

DISSENSION FROM DAVIDSON: INTERTRANSLATABILITY, COMPARABILITY AND SCHEME INDIVIDUATION

Several writers have criticized Davidson's argument by challenging Davidson's assimilation of schemes and languages, and his utilizing intertranslatability as the criterion of individuation of alternative schemes.[23] Popper, for example, argues that[24]

> It is just a dogma – a dangerous dogma – that the different frameworks are like mutually untranslatable languages. The fact is that even totally different languages (like English and Hopi, or Chinese) are not untranslatable, and that there are many Hopis or Chinese who have learnt to master English very well.

Similarly, Chris Swoyer suggests that, if we are interested in relativism, we should *not* say, with Davidson, that[25]

> . . . two groups of people possess different conceptual frameworks only if they are unable to communicate with, or translate, each other. For while we may find arguments which would lead us to conclude that this is in fact the case, few relativists would accept such a characterization. Moreover, it would rule out from the beginning many interesting questions about relativism.

Rescher, too, dissents from Davidson's view that our attribution of X's having an alternative language requires being able to translate X's utterances into our language; and more generally, from the view that intertranslatability is the key to scheme individuation.[26] These writers argue, against Davidson, that Davidson's criterion of intertranslatability is too strong, for it rules out the possibility of alternative conceptual schemes which either can be intertranslated, but are still properly thought of as alternatives, or which cannot be intertranslated but which nevertheless can be recognized as alternative schemes. I am sympathetic with the thrust of this claim (though not with its detailed presentation by some of these writers[27]). That is, I agree that Davidson's intertranslatability criterion rules out too much: granted that there cannot be alternative conceptual schemes given Davidson's criterion, it is nevertheless the case that there can be conceptual schemes which are intertranslatable, and so are not alternatives on Davidson's view, but which *are* properly viewed as alternatives and which raise (as Swoyer suggests) important epistemological questions which Davidson's analysis sweeps under the rug. If we individuate schemes not by the criterion of intertranslatability, but by that of *comparability*, we get a much different view of the possibility of alternative conceptual schemes than Davidson offers.

Consider the case of Galileo, Cremonini, and the discovery of Jovian moons.[28] According to the well-known story, Cremonini (and his cronies Libri and Magini) refused to look through Galileo's telescope at the moons, for, according to their Aristotelian conceptual scheme, those moons simply could not exist. Here there is no problem about intertranslatability. Galileo clearly understood and could formulate in his own language the Aristotelian scheme; similarly, Cremonini clearly grasped Galileo's position, and recognized the threat that it presented to his (i.e. Cremonini's) scheme. The problem is one of communication and comparative evaluation, but not of intertranslatability. Rather, as Weinert says: "Such then is the communication breakdown between Galileo and Cremonini: it is the opposition between the belief in what is observed and the belief in what the texts say there is to be observed."[29]

There *is* a problem of alternative frameworks or schemes here, a problem which raises the spectre of relativism. But it has nothing to do with intertranslatability. It has rather to do with *comparability*, i.e. with whether or not Galileo's and Cremonini's schemes can be rationally compared and assessed, whether or not we can provide grounds for rationally preferring "the Jovian moons exist" to "the Jovian moons do not exist." One might be tempted to say "just look – *that* will provide all the grounds you need." But of course such a view simply begs the question against Cremonini, who denies the relevance of observation to the issue. What is at issue here is not the existence of the Jovian moons, but the standards (telescopic observation? naked eye observation? Scripture? the writings of The Philosopher?) by which such existence claims are to be adjudicated. To take Davidson's line, and hold that Galileo and Cremonini do not have alternative schemes because their respective claims are intertranslatable, is to miss entirely the epistemological problem which the example raises. If there is no way to non-question-beggingly assess the merits of the rival standards – if standards are *in*ternal[30] to such complex webs of methodological, metaphysical and epistemological views and presuppositions as Galileo and Cremonini cling to – then we have a serious suggestion of relativism. And it seems perfectly appropriate to speak of the complex webs just referred to as alternative conceptual schemes.

Thus there is a problem with Davidson's criterion of intertranslatability. Nor is it merely a verbal quibble. We are agreed that, given Davidson's criterion, there can be no alternative schemes. We could call the complex webs something else besides "schemes," and in doing so we could preserve Davidson's conclusion. But it is crucial to see that doing so would not eliminate the possibility of conceptual or framework relativism. The complex webs are sufficient to raise this issue; neither intertranslatability nor its failure are necessary. Thus, whether or not there are alternative schemes in Davidson's sense, there are alternative schemes in a weaker sense which is nevertheless strong enough to raise the relativism issue. So Davidson's discussion fails to put the issue to rest.[31]

Scheme relativism hinges, then, not on intertranslatability, but comparability. The question is whether alternative schemes, which can be individuated by some weaker criterion than Davidson's, admit of "external" or neutral criteria by which their respective merits can be ascertained. In this sense Davidson's discussion misses the issue of relativism. The criterion of intertranslatability is too strong, for the problem of scheme relativism arises if alternative frameworks can be intertranslatable, but not externally or neutrally evaluatable. And it is this last possibility that most defenders of relativism argue for.[32]

We are back, then, in the situation in which we left Popper: acknowledging the coherence of framework relativism, but in need of an argument against it.[33] In the next section I shall offer such an argument.[34]

THE ARGUMENT WANTED

As it turns out, the argument wanted is not difficult to conceive and articulate. For a variant of the standard "reductio" argument against epistemological relativism in general works perfectly well when aimed at framework relativism in particular. Quine[35] puts the point forcefully in terms of cultural relativism:

> Truth, says the cultural relativist, is culture-bound. But if it were, then he, within his own culture, ought to see his own culture-bound truth as absolute. He cannot proclaim cultural relativism without rising above it, and he cannot rise above it without giving it up.

Substitute "framework" for "cultural" and "framework-bound" for "culture-bound" in this passage, and the argument wanted is at hand. The framework relativist, if correct, would be unable to recognize the equal status of alternative frameworks, for she would not be able to transcend her own. She would thus regard her own framework as absolute. But she does recognize alternative frameworks. So it cannot be the case that her framework constitutes the bound on her thought that is essential to the framework relativist position. As Quine puts it, she cannot proclaim framework relativism without rising above it; and she cannot rise above it without giving it up.[36] The recognition of the equal epistemological status of (and indeed of the independent existence of) alternative frameworks, which is necessary for framework relativism, necessitates as well the rejection of framework-boundedness, which is the central component of the framework relativist position. Thus framework relativism cannot proclaim itself, or even recognize itself, without defeating itself.

Framework relativism is, in the end, simply a species of epistemological relativism. As I argued in Chapter 1, the genus is open, generally, to an updated and revised version of the classical *reductio* argument against epistemological relativism; so, *a fortiori*, is the species. It would be inappropriate to repeat here the detailed account of the revised *reductio* argument developed in the last chapter, but perhaps a brief review is in order. I offer one in what follows, applying the general anti-relativist argument to framework relativism in particular. First, the framework relativist must, in order to join the issue with the non-relativist, defend framework relativism non-relativistically. To "defend" framework relativism relativistically (i.e. "according to my framework, framework relativism is true (correct, warranted, etc.)") is to fail to *defend* it, since the non-relativist is appropriately unimpressed with such framework-bound claims. But to defend framework-relativism non-relativistically is to give it up, since to defend it in this way is to acknowledge the legitimacy of framework-neutral criteria of assessment of claims, which is precisely what the framework relativist must deny. Thus to defend framework relativism relativistically is to fail to defend it; to defend it non-relativistically is to give it up. Thus the defense of framework relativism insures its defeat. Thus framework relativism is self-defeating. (The relation between this argument and the Quinean one offered above is I hope clear.)

The self-defeat and so incoherence of framework relativism can also be seen in a second way. If framework relativism is right (or true, or correct, or epistemically worthy, etc.), then no transcendence of frameworks is possible. But if no such transcendence is possible, then neither is any non-question-begging evaluation of the framework relativist thesis (or the framework from which it stems). Consequently, it is impossible in principle to defend framework relativism as right, for the very notion of rightness is given up by the framework relativist. But then framework relativism cannot be right. Rightness (or epistemic worthiness) presupposes neutral criteria or a neutral "meta-framework" from which rival frameworks, and the claim that framework relativism is right, can be judged. But this is precisely what the framework relativist denies. Thus by this argumentative route as well we reach the conclusion that framework relativism is self-defeating. These several considerations provide, I believe, what Popper and Davidson unsuccessfully tried to provide: considerations which show that framework relativism is untenable.

I conclude that considerations of schemes, frameworks and the like in the end offer no comfort to the epistemological relativist. Like other classical forms of epistemological relativism, framework relativism cannot be maintained – or if it can, the argument which would show this has yet to be developed.[37]

We have up to this point been treating general epistemological considerations regarding relativism. In the next four chapters our focus shifts to relativism's manifestation in recent philosophy of science. The discussion begins, in the next chapter, with the pivotal work of Kuhn.

PART II

RELATIVISM AND THE PHILOSOPHY OF SCIENCE

CHAPTER 3

KUHN AND RELATIVISM: IS HE OR ISN'T HE?

Thomas S. Kuhn's *The Structure of Scientific Revolutions* is perhaps the single most influential work in the "new", i.e. post-positivist, philosophy of science. Kuhn is widely regarded as having offered a relativistic conception of scientific knowledge. Yet he has disavowed relativism, and has made several attempts to clarify his position regarding relativism and related issues. As we shall see, relativism is for Kuhn bound up with a host of such issues, especially those concerning the rationality and objectivity of science, and the "incommensurability" of rival scientific theories. Consequently, we must consider the entire Kuhnian conception of science in order to come to grips with Kuhn's commitment to relativism. Thus Kuhn's general portrayal of science constitutes the focus of this chapter. Our tasks are to determine the extent of Kuhn's commitment to a relativistic epistemology of science, to sort out the complex tangle of charge and counter-charge between Kuhn and his critics, and to assess the power of the broad Kuhnian picture of science, especially regarding relativism, the objectivity of scientific knowledge, and the rationality of scientific methodology. We begin with an exposition of Kuhn's book.

A. INITIAL EXPOSITION

(1) *Paradigms and Normal Science*

The most fundamental notion of Kuhn's conception of the dynamics of science is that of *paradigm*. Paradigms are "universally recognized scientific achievements that for a time provide model problems and solutions to a community of practitioners."[1] Paradigms, then, are scientific achievements that guide the research of some group of scientists. They are "accepted examples of actual scientific practice – examples which include law, theory, application, and instrumentation together – [which] provide models from which spring particular coherent traditions of scientific research."[2] Paradigms share "two essential characteristics.

Their achievement is sufficiently unprecedented to attract an enduring group of adherents away from competing modes of scientific activity. Simultaneously, it is sufficiently open-ended to leave all sorts of problems for the redefined group of practitioners to resolve."[3] The open-ended character of these paradigmatic achievements is crucial to Kuhn's view, for it allows for scientific work to continue under a paradigm. Such work Kuhn labels *normal science*. (This contrasts, as we shall see, with Kuhn's notion of *revolutionary science* – research carried on which is not unambiguously connected with a particular paradigm.) Normal science is "research firmly based upon one or more past scientific achievements, achievements that some particular scientific community acknowledges for a time as supplying the foundation for its further practice."[4] The notion of "normal science" is thus very close to the notion of "paradigm" – normal science is simply research carried on under the guidance of a particular paradigm; and, likewise, a paradigm is that which characterizes normal scientific research.

Scientists who share a paradigm share more than a unique scientific achievement which guides their research. "Men whose research is based on shared paradigms are committed to the same rules and standards for scientific practice. That commitment and the apparent consensus it produces are prerequisites for normal science, i.e., for the genesis and continuation of a particular research tradition."[5] Paradigms guide normal science, then, by specifying "rules and standards" for conducting and appraising such research.

Kuhn construes normal science as "puzzle-solving" rather than as "problem-solving." This is because, for the typical normal science activity, there are rules which guide the solving process, and there is the assurance of an answer – these rules and assurances both provided by the paradigm under which the research is carried on.[6] (Bona fide problem solving, for Kuhn, is not so guaranteed of an answer, and is not always guided by clearcut rules and procedures. Such problem solving belongs to revolutionary science, where the absence of a guiding paradigm accounts for these difficulties.) Normal science is "an enterprise that . . . aims to refine, extend, and articulate a paradigm that is already in existence."[7] Normal science does not seek novelty. It is, rather, "a strenuous and devoted attempt to force nature into the conceptual boxes supplied by professional education."[8] Normal science may be seen, consequently, as a servant to its paradigm – it is research so thoroughly dominated by a paradigm that it seeks not simply to discover interesting new truths about the world, to uncover facts about nature, but rather strives to make its paradigm "look good." That is to say, normal science seeks to find ways to construe nature so that it can be accounted for by its paradigm. This is the upshot of Kuhn's conception of normal science as "refining, extending, and articulating" a paradigm, and his comments about "forcing nature into the conceptual boxes supplied by professional education" – such education being, of course, the administering of the dominant paradigm.

How does a scientific achievement gain the status of a paradigm? Kuhn argues that such status is in large part due to actual, and potential, problem solving ability. "Paradigms gain their status because they are more successful than their competitors in solving a few problems that the group of practitioners has come to recognize as acute . . ."[9] This problem-solving ability is often not an established fact, however; it is rather a *potential* problem-solving capability which offers merely the promise of success – a promise to be fulfilled by normal science. "The success of a paradigm . . . is at the start largely a promise of success discoverable in selected and still incomplete examples. Normal science consists in the actualization of that promise, an actualization achieved by extending the knowledge of those facts that the paradigm displays as particularly revealing, by increasing the extent of the match between those facts and the paradigm's predictions, and by further articulation of the paradigm itself."[10] A paradigm, further, must be "able to guide the whole group's research."[11]

Once a scientific achievement gains the status of a paradigm, how it loses that status – that is, how it is challenged and replaced by competing paradigms – is a crucial portion of Kuhn's conception of the dynamics of science. It is to this process that we now turn.

(2) *Anomaly and Crisis*

A successful, well-entrenched paradigm is, it might seem, impossible to dethrone. Normal science is conducted in accordance with its rules and standards; the goal of normal science is to "articulate" the paradigm and "force nature into the conceptual boxes" it dictates; and young scientists are trained in terms of it. How, then, once a paradigm gains favor, can it fall from grace?

This can only happen, according to Kuhn, when normal research conducted under a paradigm discovers *anomalies*. "Discovery commences with the awareness of anomaly, i.e., with the recognition that nature has somehow violated the paradigm-induced expectations that govern normal science."[12] An anomaly, then, is a circumstance that can't be rendered intelligible by, or explained in terms of, the paradigm whose normal scientific research uncovered it. It is a normal scientific "puzzle" which cannot be solved and so, for Kuhn, ceases to be a puzzle and becomes a problem. It is anomaly that precipitates scientific revolutions:

> Sometimes a normal problem, one that ought to be solvable by known rules and procedures, resists the reiterated onslaught of the ablest members of the group within whose competence it falls. On other occasions a piece of equipment designed and constructed for the purpose of normal research fails to perform in the anticipated manner, revealing an anomaly that cannot, despite repeated effort, be aligned with professional expectation. In these and other ways besides, normal science repeatedly goes astray. And when it does – when, that is, the profession can no longer evade anomalies that subvert the existing tradition of scientific practice . . . then begin the extraordinary investigations that lead the profession at last to a new set of commitments, a new basis for the practice of science. The extraordinary episodes in which that shift of professional commitment occurs are the ones known in this essay as scientific revolutions. They are the tradition-shattering complements to the tradition-bound activity of normal science.[13]

When normal science uncovers anomalies that resist solutions, the paradigm which governs such research is said to be in *crisis*. When in crisis, "scientists take a different attitude toward existing paradigms, and the nature of their research changes accordingly. The proliferation of competing articulations, the willingness to try anything, the expression of explicit discontent, the recourse to philosophy and to debate over fundamentals, all these are symptoms of a transition from normal to revolutionary research."[14] It is only in such states of crisis, precipitated by anomaly, that competing paradigms are advanced and paradigm debate begins: "Novel theory emerge[s] only after a pronounced failure in the normal problem-solving activity."[15] Once in crisis, however, a paradigm is not dropped; it is retained until a new paradigm arrives on the scene which can make sense of the anomalies which precipitated the crisis: "Once it has achieved the status of paradigm, a scientific theory is declared invalid only if an alternate candidate is available to take its place."[16]

So, while a well-entrenched paradigm is indeed difficult to dislodge – "In science . . . novelty emerges only with difficulty, manifested by resistance, against a background provided by expectation"[17] – no paradigm is permanently entrenched. The existence of anomaly, if severe enough, is sufficient to begin full-scale examination of the paradigm in question. When such questioning is undertaken, the paradigm is in crisis, and the search begins for alternative paradigms, paradigms which at least promise to solve the anomalies that led to the crisis. Once potential replacement paradigms are found, debate ensues concerning the relative merits of the old and new paradigms. Such debate is called *paradigm debate*. Should a new paradigm replace the old one, we have, for Kuhn, scientific revolution. The nature of paradigm debate is a very important part of Kuhn's views, and, indeed, his most far-reaching and radical opinions center on the nature of such debate. We shall next consider, therefore, the nature of scientific revolution and paradigm comparison.

(3) *Revolutionary Science and the Incommensurability of Competing Paradigms*

Scientific revolutions are, for Kuhn, indeed revolutionary – the change from one paradigm to the next is discontinuous and non-cumulative. Kuhn sees a new paradigm replacing an older one, rather than following from it. The old and new paradigms are "irreconcilable"[18] and incompatible:" . . . scientific revolutions are here taken to be those non-cumulative developmental episodes in which an older paradigm is replaced in whole or in part by an incompatible new one."[19] Kuhn's departure here from the traditional view of science as a cumulative process[20] is striking. His departure from this traditional view stems from his conception of paradigms and paradigms debate. Kuhn conceives of competing paradigms as not only incompatible but *incommensurable* – incapable of being compared, or "measured against," a neutral standard: "The normal-scientific tradition that emerges from a scientific revolution is not only incompatible but often actually incommensurable with that which has gone before."[21] The reasons for the incommensurability of competing paradigms, and the implications of their incommensurability, are explored by Kuhn in the following passage:

> In learning a paradigm the scientist acquires theory, methods, and standards together, usually in an inextricable mixture. Therefore when paradigms change, there are usually significant shifts in the criteria determining the legitimacy both of problems and of proposed solutions.
>
> That observation . . . provides our first explicit indication of why the choice between competing paradigms regularly raises questions that cannot be resolved by the criteria of

normal science . . . They [scientists of different paradigms] will inevitably talk through each other when debating the relative merits of their respective paradigms. In the partially circular arguments that regularly result, each paradigm will be shown to satisfy more or less the criteria that it dictates for itself and to fall short of a few of those dictated by its opponent.[22]

So, Kuhn argues that competing paradigms are incommensurable because paradigms include not only theories, but standards and criteria of evaluation as well. Competing paradigms are thus incommensurable – cannot be measured against a neutral standard – because each paradigm possesses a built-in non-neutral standard of its own. The implications of this view of Kuhn's are staggering, for such a view implies that neutral, objective debate between the two competing paradigms is impossible. This is perhaps the most controversial and radical claim connected with Kuhn's philosophy of science; consequently, we shall now consider the claim in more detail.

Is paradigm debate necessarily non-neutral – that is, subjective, arbitrary and/or irrational? Despite occasional protestation to the contrary, it seems that Kuhn is committed to an affirmative answer to this question. He likens the scientist's decision to adopt a new paradigm to the "gestalt switch":

. . . normal science ultimately leads only to the recognition of anomalies and to crises. And these are terminated, not by deliberation and interpretation, but by a relatively sudden unstructured event like the gestalt switch. Scientists often speak of the "scales falling from the eyes" or of the "lightning flash" that "inundates" a previously obscure puzzle . . .[23]

or again: "Just because it is a transition between incommensurables, the transition between competing paradigms cannot be made a step at a time, forced by logic and neutral experience. Like the gestalt switch, it must occur all at once . . . or not at all.[24]

Kuhn construes paradigm debate as irrational in other ways, besides his contention that such transitions are like gestalt switches. He argues more generally that the paradigm-bound nature of criteria of evaluation of paradigms makes paradigm debate circular, and so merely a matter of persuasion: "When paradigms enter, as they must, into a debate about paradigm choice, their role is necessarily circular. Each group uses its own paradigm to argue in that paradigm's defense."[25] Furthermore, ". . . the status of the circular debate is only that of persuasion. It cannot be made logically or even probabilistically compelling for those who refuse to step into the circle."[26] Kuhn also argues that "The proponents of competing paradigms are always at least slightly at cross-purposes"[27] and that ". . . they are bound partly to talk through each other."[28] He characterizes paradigm debate as "communication across the revolutionary divide," which, he claims, "is inevitably partial."[29] Finally, Kuhn labels three aspects of the incommensurability of competing paradigms, which serve again to construe paradigm debate as irrational: First, "the proponents of competing paradigms will often disagree about the list of problems that any candidate for paradigm must resolve. Their standards or their definitions of science are not the same."[30] Second, "within the new paradigm, old terms, concepts, and experiments fall into new relationships one with the other. The inevitable result is what we must call, though the term is not quite right, a misunderstanding between the two competing schools."[31] Third, ". . . the proponents of competing paradigms practice their trades in different worlds."[32]

Despite the fact that Kuhn characterizes the acceptance of a new paradigm as a gestalt switch, which can hardly be construed as a decision (one doesn't *decide* to form a new gestalt), Kuhn considers the nature of the scientist's decision to adopt a new paradigm. Here he adopts a religious metaphor, and claims that: "The transfer of allegiance from paradigm to paradigm is a conversion experience that cannot be forced."[33] He claims that: "A decision of that kind [to adopt a new paradigm] can only be made on faith"[34] and that: "There must be a basis, though it need be neither rational nor ultimately correct for faith in the particular candidate chosen."[35]

Kuhn begins to give a more tangible reason for the decision to adopt a new paradigm when he considers the *problem-solving ability* of the new candidate for paradigm: "Probably the single most prevalent claim advanced by the proponents of a new paradigm is that they can solve the problems that have led the old one to a crisis."[36] He expands on the point, but then seemingly denies that problem-solving ability can serve as a basis for paradigm choice in the following passage:

> . . . the new candidate must seem to resolve some outstanding and generally recognized problem that can be met in no other way. Second, the new paradigm must promise to preserve a relatively large part of the concrete problem-solving ability that has accrued to science through its predecessors . . .
>
> To say this is not to suggest that the ability to solve problems is either the unique or an unequivocal basis for paradigm choice. We have already noted many reasons why there can be no criterion of that sort.[37]

So in the end, the decision to accept a new paradigm is more like a religious than a rational act.

To sum up this section: Kuhn argues that scientific change is not cumulative, but that a new paradigm is incompatible and even incommensurable with its predecessor. Paradigms are incommensurable mainly because they are "theories, methods, and standards" all in one – that is to say, they include criteria of evaluation of theory as well as theory itself. Incommensurability thus stems from the fact that two competing paradigms have conflicting criteria of evaluation. Incommensurability has a profound implication for paradigm debate: namely, it seems to render such debate irrational. Such an irrationalist portrayal of theory choice makes scientific knowledge relative as well, since judgments of factual and theoretical adequacy are on this picture relative to the incompatible criteria of evaluation fostered by rival paradigms.[38]

B. THE EARLY CRITICISMS

It is perhaps not surprising that many have found this irrationalist strain in Kuhn objectionable, and have sought to criticize Kuhn on this point. Note that those aspects of Kuhn's initial formulation which suggest that paradigm-choice is irrational (henceforth the "irrationality thesis") stem largely from his contention that paradigms are incommensurable. Incommensurability – the inability of competing paradigms to be directly compared, or judged according to a neutral standard – stems from Kuhn's contention that a paradigm contains its own criteria of evaluation, as well as laws and methods of application. During paradigm debate, according to Kuhn, competing paradigms are evaluated according to paradigm-bound criteria of evaluation; because of incommensurability,[39] there are no paradigm-neutral criteria of evaluation. Since there are no paradigm-neutral criteria of evaluation, paradigm debate can rely on no objective criteria of evaluation of paradigms; hence paradigm debate is irrational. The irrationality thesis thus rests on incommensurability, which in turn rests on the paradigm-bound nature of criteria of evaluation of paradigms. The irrationality thesis is therefore open to criticism through criticism of the paradigm-bound conception of criteria of evaluation of paradigms.[40] This is, in fact, a major focus of some of the critics of the irrationality thesis. Scheffler, Shapere and Kordig, for example, have attacked the irrationality thesis in just this way – by criticizing Kuhn's arguments for the paradigm-bound nature of criteria of evaluation of paradigms. Kuhn, it will be recalled, defines paradigms as including methods and criteria of evaluation as well as theory, so that proponents of competing paradigms assess paradigms according to different, incompatible sets of criteria of evaluation of paradigms. As Scheffler describes Kuhn's position:

> . . . to accept a paradigm [according to Kuhn] is to accept not only theory and methods, but also governing standards or criteria which serve to justify the paradigm as against its rivals, in the eyes of its proponents. Paradigm differences are thus inevitably reflected upward, in criterial differences at the second level. It follows that each paradigm is, in effect, inevitably self-justifying, and that paradigm debate must fail of objectivity.[41]

Scheffler criticizes this view of Kuhn's by arguing that it blurs a crucial distinction by failing to distinguish between criteria *internal* to a paradigm and criteria *external* to a paradigm. Kuhn is perhaps correct that each paradigm sets criteria such that within a paradigm, certain pieces of evidence or experimentation can only be interpreted or evaluated according to the criteria set forth by the paradigm; however, he has failed to produce any plausible reason for thinking that external criteria – this is, criteria according to which paradigms are themselves judged – must be included in a paradigm. Paradigm debate is more properly construed as a second-order, or meta-activity, and at this higher level of evaluation independent criteria are operative. The distinction between internal and external criteria thus renders gratuitous Kuhn's claim that paradigms are "self justifying" – for in light of the internal/external distinction, it must be concluded that competing paradigms are judged according to external criteria, hence not according to criteria which stem from and thus favor one of the paradigms being judged as against its rivals. Kordig, in agreement with Scheffler, writes that:

> Kuhn has failed to show that paradigm differences imply evaluative differences at a second, or meta, level; he has not shown that the sharing of second-order criteria by enthusiasts of rival paradigms is impossible. He has, therefore, not demonstrated that scientific transitions consist only in non-cumulative persuasions and conversions.[42]

By bringing to light this crucial objection to Kuhn's argument for the paradigm-bound nature of criteria of evaluation of paradigms, Scheffler and Kordig have cast great doubt on Kuhn's incommensurability claim – for if second order principles are operative in paradigm-evaluation, there need be no incommensurability between competing paradigms, since they are capable of being evaluated according to these second-order principles. Denial of incommensurability, in turn, renders suspect the irrationality thesis, since that thesis, as we saw earlier, depends in large part on incommensurability. Moreover, with incommensurability goes relativism, because if second order principles function objectively in paradigm evaluation, then judgments of theoretical adequacy are not relative to rival paradigms. Kuhnian irrationalism and relativism are thus severely challenged by this challenge to Kuhn's incommensurability thesis.

C. KUHN'S INITIAL RESPONSE

Kuhn initially responded to these criticisms not by defending the irrationality and relativism theses, but by denying them and complaining that his critics misinterpret his position.[43] He writes:

> My critics respond to my views . . . with charges of irrationality, relativism, and the defense of mob rule. These are all labels which I categorically reject.[44]

Kuhn claims that his critics err when they report that he believes that

> the proponents of incommensurable theories cannot communicate with each other at all; as a result, in a debate over theory-choice there can be no recourse to *good* reasons; instead theory must be chosen for reasons that are ultimately personal and subjective; some sort of mystical apperception is responsible for the decision actually reached.[45]

If this is misinterpretation, what is the correct interpretation of Kuhn's views? Kuhn clarifies his position by claiming, in effect, that there is no algorithm, no rules, for theory-choice. He argues that there are never deductive proofs which necessitate the choice of one theory over another, and that debates over theory-choice can never end with a conclusive logical verdict.[46] This is, as Kuhn recognizes, hardly an unfamiliar thesis; indeed, it is a proposition generally agreed to by all parties in the debate. Why, then, if this is all Kuhn meant to say concerning paradigm-choice, all the rhetoric about "conversion", "gestalt switches," and the like during scientific revolutions? Kuhn attempts to clarify his position through the introduction of "values":

> Nothing about that relatively familiar thesis [i.e. that there is no algorithm for theory choice] implies either that there are no *good reasons* for being persuaded or that these reasons are not ultimately decisive for the group. Nor does it even imply that the reasons for choice are different from those usually listed by philosophers of science: accuracy, simplicity, fruitfulness, and the like. What it should suggest, however, is that such reasons function as *values* and that they can thus be differently applied, individually and collectively, by men who concur in honoring them. If two men disagree, for example, about the

relative fruitfulness of their theories, or if they agree about that but disagree about the relative importance of fruitfulness and say, scope in reaching a choice, neither can be convicted of a mistake. Nor is either being unscientific. There is no neutral algorithm for theory-choice, no systematic decision procedure which, properly applied, must lead each individual in the group to the same decision.[47]

This paragraph of Kuhn's seems unexceptionable (disregarding for a moment the mention of "values"), and is a far cry from the irrationality thesis. There is no mention here of paradigm-bound criteria of evaluation; no mention of incommensurability. However, there is a question about the notion of "values" Kuhn introduces, for the relationship between values and the irrationality thesis needs to be spelled out. It also needs to be pointed out that Kuhn, in this passage, allows himself recourse to "good reasons" during paradigm-choice; whether or not Kuhn is entitled to do so will also be considered presently. For now, I wish to stress the non-radicalness of the cited paragraph, and the extent to which it softens the claims of Kuhn's original presentation of the irrationality thesis. Let us now turn our attention to the remainder of Kuhn's remarks. He continues the passage:

. . . To understand why science develops as it does, one need not unravel the details of biography and personality that lead each individual to a particular choice, though that topic has vast fascination. What one must understand, however, is the manner in which a particular set of shared values interacts with the particular experiences shared by a community of specialists to ensure that most members of the group will ultimately find one set of arguments rather than another decisive.

That process is persuasion, but it presents a deeper problem. Two men who perceive the same situation differently but nevertheless employ the same vocabulary in its discussion must be using words differently. They speak, that is, from what I have called incommensurable viewpoints.[48]

Here Kuhn has brought back into the discussion incommensurability, and consequently, the irrationality thesis. The tension between the two cited passages runs throughout Kuhn's initial responses, and is in large part the reason his critics fail (properly) to be quieted by these responses. He continually tries both to retract the more radical claims of his earlier writing, and at the same time to maintain them under different guises. Besides the just-cited re-introduction of incommensurability, Kuhn's newer views depend on the notions of good reasons and values, notions which are not clearly reconcilable with his original presentation:

What I am denying then is neither the existence of good reasons nor that these reasons are of the sort usually described. I am, however, insisting that such reasons constitute values to be used in making choices rather than rules of choice.[49]

These remarks make Kuhn's overall position difficult to make sense of. It is not clear that these just-cited passages concerning the role of values, and the existence of good reasons for theory choice, are consistent with his original presentation of incommensurability. Consider, for example, Kuhn's emphasis on one particular characteristic of values: they "may be shared by men who differ in their application,"[50] so that two scientists may share the value of, say, simplicity, yet differ as to the simplicity of their competing theories, or the importance of simplicity relative to another value, say scope. This is unexceptionable, and is little more than elaboration of the "no algorithm for theory choice" theme. However, it is at odds with incommensurability, if Kuhn means to suggest that the two scientists can deliberate meaningfully on the question of the simplicity or scope of their respective theories. If Kuhn is willing to allow for such deliberation, which some of the just-cited remarks seem to indicate, his comments on values are clearly inconsistent with his maintenance of incommensurability, for incommensurability guarantees the impossibility of such deliberation. If, on the other hand, Kuhn is not willing to allow for such deliberation, it is difficult to see how the introduction of "values" into the discussion enlarges on, or clarifies, the incommensurability claim. The status of values, then, needs to be clarified. As Shapere puts the problem in his review of Kuhn's early replies to his critics, "if there are . . . no constraints on what one can assert in the name of 'values'," Kuhn's viewpoint is "as relativistic, as antirationalistic, as ever."[51] If there *are* constraints on what one can assert in the name of values, on the other hand, it is difficult to see how Kuhn can reconcile those constraints with his incommensurability claim – for incommensurability demands the absence of such constraints.

This ambiguity permeates Kuhn's discussion of *good reasons* for theory choice as well. Kuhn insists that there are good reasons for theory choice, but that these reasons "constitute values." Because Kuhn's account of values is unclear, his account of reasons is as well. For if there are no constraints on what one can assert in the name of values, it is difficult to see the sense in which such assertions can be taken to constitute objective reasons for choosing one theory over another. Moreover, given incommensurability, it is difficult to make any sense of the term "good reason" as it applies to paradigm debate, since, as Shapere puts it, "the very notion of a 'good reason' has been made paradigm-dependent."[52]

Kuhn's introduction of "values" and "good reasons,"therefore, seem unhelpful in clearing up the debate over the irrationality thesis. As we have seen, Kuhn has explicitly denied the thesis;[53] explicitly maintained incommensurability, from which the thesis follows;[54] and introduced into the discussion notions which seem to do nothing but further cloud his view. If we take Kuhn's denial of the irrationality thesis seriously, he may then be a proponent of objectivity in science – thus agreeing with his critics. However, it is difficult to take that denial seriously, in light of his reluctance to give up the incommensurability thesis, his hedging over the status of reasons, and his continued descriptions of theory choices as gestalt switches and conversion experiences:

> The conversion experience that I have likened to a gestalt switch remains, therefore, at the heart of the revolutionary process. Good reasons for choice provide motives for conversion and a climate in which it is more likely to occur. Translation may, in addition, provide points of entry for the neural programming that, however inscrutable at this time, must underlie conversion. But neither good reasons nor translation constitute conversion, and it is that process we must explicate in order to understand an essential sort of scientific change.[55]

As the above passage indicates, Kuhn "seems unwilling to abandon 'incommensurability' while trying, unsuccessfully, to assert that communication and comparison are possible."[56] Kuhn seems to want it both ways: he wants to maintain incommensurability (and so irrationality), yet deny irrationality and allow for communication between proponents of competing paradigms (thus giving up incommensurability). It is clear, I hope, that Kuhn cannot have it both ways. His maintenance of incommensurability vitiates his denial of the irrationality thesis and of relativism.

D. KUHN'S FURTHER RESPONSE

Thus far, I have been reviewing the early debate between Kuhn and Scheffler, Shapere and Kordig over the objectivity of scientific theory choice. We have seen that Kuhn's early responses to these critics have been unconvincing, and have done little to enhance the plausibility of his original presentation of incommensurability. In this section of the present chapter, I shall consider three more recent attempts by Kuhn to answer his critics on the interrelated issues of objectivity, irrationality, relativism and incommensurability. I will argue that Kuhn's further responses fail either (1) to establish a tenable irrationality or incommensurability or relativism thesis; or (2) to eliminate the tension, illustrated in the previous section of this chapter, between Kuhn's wanting to maintain a strong incommensurability claim, on the one hand, and his wanting to weaken or eliminate such a claim, on the other.

(1) *Comparability*

At least part of the reason that incommensurability leads to arbitrary, and so irrational, choice among competing scientific theories is that incommensurability seems to preclude the very possibility of objective comparison of rival theories. This is clearly the thrust of Kuhn's *The Structure of Scientific Revolutions*: incommensurability precludes comparability. This seems to raise a major problem for Kuhn, for, as Scheffler has noted, if incommensurable theories cannot be compared, they cannot be recognized as rivals. But if theories cannot be recognized as rivals, they cannot be said to compete. Thus the very possibility of theory *choice* from among rival competitors is undermined. But of course this result is absurd. We do, for example, recognize Newtonian and relativistic mechanics as competitors. Therefore, there must be something very wrong with the "incommensurability precludes comparability" thesis.

The thesis, however, is quite right: given incommensurability, incomparability follows. What is wrong is that the incommensurability thesis has not been shown to be true. Moreover, the fact that we do recognize rival theories as rivals constitutes a decisive objection to incommensurability. For to recognize rival theories as rival is to deny their incommensurability at the meta-level. The proper conclusion to be drawn from Scheffler's point is not, therefore, that incommensurability is compatible with comparability, and so is no threat to the possibility of rational theory choice.[57] Rather, we must conclude that the fact that comparison is possible raises grave doubts about the incommensurability thesis. Object-level (or observation-level) incommensurability does not of itself preclude comparability; object-level incommensurability does not imply meta-level incommensurability, so that there is no reason to deny that objective theory comparison is possible. But total (i.e. object- and meta-level) incommensurability does indeed preclude comparability, and so guarantees the irrationality of theory choice – for how can choice be rational if the candidate theories cannot be compared, or even recognized as alternatives?[58]

Kuhn has, in a more recent paper,[59] addressed this issue of the relationship between incommensurability and comparability. Unfortunately, Kuhn's discussion both fails to come to grips with our reformulation of Scheffler's criticisms of the original "incommensurability precludes comparability" argument, and fails to clarify or refine his position concerning incommensurability. He begins his discussion by seemingly granting Scheffler's point that object-level incommensurability does not preclude comparability:

> Most readers of my text have supposed that when I spoke of theories as incommensurable, I meant that they could not be compared. But 'incommensurability' is a term borrowed from mathematics, and there it has no such implication. The hypotenuse of an isosceles right triangle is incommensurable with its side, but the two can be compared to any required degree of precision. What is lacking is not comparability but a unit of length in terms of which both can be measured directly and exactly.[60]

These remarks, it seems, grant Scheffler's point that object-level incommensurability does not preclude comparability – and so, that objective comparisons of rival scientific theories is possible. Taken by themselves, these remarks constitute, if not a complete repudiation, at least a substantial weakening of Kuhn's original position on incommensurability and comparability. Unfortunately, however, Kuhn continues his comments here in a way which renders his overall position confusing. He continues the just-cited passage as follows:

> In applying the term "incommensurability" to theories, I had intended only to insist that there was no common language within which both could be fully expressed and which could therefore be used in a point-by-point comparison between them.[61]

This last sentence reintroduces the incomparability the earlier remarks deny. For unless there is a substantive difference between "comparison" and "point-by-point comparison," Kuhn is saying that incommensurable paradigms can be compared, but not compared "point-by-point." This is equivalent to saying that they can be compared, but not compared, which does little to illuminate Kuhn's position. Moreover, in a footnote at this juncture in his paper, Kuhn pointedly returns to his original "incommensurability precludes comparability" position, as well as his original view that competing theories are indeed incommensurable, so that objective, rational theory choice is impossible:

When I first made use of the term "incommensurability," I conceived the hypothetical neutral language as one in which any theory at all might be described. Since then I have recognized that comparison requires only a language neutral with respect to the two theories at issue, but I doubt that anything of even that limited neutrality can be designed.[62]

What can be made of this? In the space of a single paragraph and footnote, Kuhn has allowed for the possibility of neutrally comparing rival scientific theories (thus repudiating his original incommensurability thesis); doubted the possibility of such neutral comparison; and denied the possibility of "point-by-point" neutral comparison (while failing to explain the alleged difference between "point-by-point" comparison and comparison *simpliciter*). It seems clear that these recent comments by Kuhn do little to clarify or make plausible his position concerning incommensurability, comparability, and the objectivity of scientific theory choice. As we saw in the previous section of this chapter concerning Kuhn's initial response to his critics, this later response does little to eliminate the tension between Kuhn's wanting both to maintain a strong incommensurability claim, and simultaneously to allow for rational and objective theory choice in science. And, again as we saw earlier, he cannot have it both ways. Moreover, it must be emphasized that the fact of comparability constitutes a powerful argument against incommensurability.

(2) *"Objectivity, Value Judgment, and Theory Choice"*

Kuhn's most recent systematic treatment of the issues we have been considering is his "Objectivity, Value Judgment, and Theory Choice."[63] However, here again his arguments do little to increase the plausibility of his position.

Kuhn's presentation is largely concerned to further develop two themes already familiar to the reader: the "no algorithm for theory choice" theme, and the "values" theme. These we have already discussed at some length, and Kuhn adds little that requires still further inspection of this much-trodden ground. Kuhn does add a new dimension to the discussion, however. In the course of his discussion of these themes, he suggests that his critics rely on an untenable distinction – the distinction, first drawn by Hans Reichenbach,[64] between the contexts of discovery and justification. It is worthwhile, in view of the importance of this distinction, to consider Kuhn's arguments here in some detail. Kuhn doubts "whether or not th[e] invocation of the distinction between contexts of discovery and of justification provides even a plausible and useful idealization"[65] of scientific activity, and offers three reasons for this doubt.

(a) The first reason centers on what Kuhn calls the "context of pedagogy," which is significantly distinct from the more familiar contexts. An important feature of science pedagogy is the tendency, Kuhn argues, to oversimplify the decisions, including decisions concerning theory choice, that scientists actually make. By relying on "famous crucial experiments" for exemplification of scientific decisions, science textbooks and philosophy of science textbooks make scientific decisions look too easy, for crucial experiments "are paradigms of good reason for scientific choice; they illustrate the most effective of all the sorts of argument which could be available to a scientist uncertain which of two theories to follow."[66] If crucial experiments make scientific decisions easy, Kuhn argues, they are also typically not available for scientists to consider when choosing between rival theories: "By the time they [the crucial experiments] were performed no scientist still needed to be convinced of the validity of the theory their outcome is now used to demonstrate. Those decisions had long since been made on the basis of significantly more equivocal evidence."[67] Therefore, to the (considerable) extent that philosophers of science take crucial experiments to be typical reason-providers for scientific decisions, they have in mind a highly misleading picture of actual scientific decision making: "The exemplary crucial experiments to which philosophers again and again refer . . . provides needed economy to science pedagogy, but they scarcely illuminate the character of the choices that scientists are called upon to make."[68]

Kuhn's historical claim, that crucial experiments do not (typically) guide scientific decisions, I will not challenge here. For even if Kuhn is correct, no ramifications for the distinction between discovery and justification follow. To see this, we need only note that scientific decisions are one thing, the justification of scientific decisions another. Scientific decision-making is, of course, a crucial sort of scientific activity; and if we seek to understand the scientific enterprise in all its aspects, we shall certainly want to study actual scientific decision-making and theory-choosing. Still, once we have a thorough account of scientist S's decision to adopt theory T_1 instead of T_2, we may well want to ask the further question: was S's decision justified? For what we are after, in the context of justification, is an account of the justificatory force of the scientist's reasons for adopting T_1; an account that enlightens us as to why those reasons are *good* reasons. Factors accounting for the scientist's decision to adopt T_1 need offer no such justificatory account – though they may well settle the question of why the scientist in fact made the decision she did. The question the context of justification is concerned to answer is not "why did X choose T_1?", but rather "was X's choice of T_1 a good one?" Given the legitimacy of this latter question, we must recognize that the "context of decision" is not equivalent to the context of justification – even though, as Kuhn correctly points out, in the case of the crucial experiment the prime factor in the scientist's decision (the results of the crucial experiment) may well be the factor that justifies that decision. Thus, we can agree with Kuhn that the discovery/justification distinction does little to illuminate actual scientific decision-making, for ceding that point is no way casts doubt on the legitimacy of the distinction between the contexts.

(b) A second, related, reason for rejecting the distinction, Kuhn argues, is that typical philosophical illustrations of scientific choice distort "by making choice [seem] totally unproblematic."[69] Typically, Kuhn claims, philosophical illustrations of scientific choice situations present these situations in ways which indicate that one theory, and not the other, is clearly and obviously to be preferred: for example, in accounts of the choice between the oxygen and phlogiston theories, oxygen's advantages over phlogiston typically are presented as unequivocally clear and, seemingly, decisive; however, the advantages the phlogiston theory enjoyed over the oxygen theory are rarely taken into account. Thus, the typical portrayal of the scientist's decision to adopt the oxygen theory instead of the phlogiston theory (or, e.g., the Copernican theory instead of Ptolemy's) misleads in that it fails to show that, in typical decision contexts, "there are always at least some good reasons for each possible choice."[70]

Thus far, Kuhn's point seems clear enough – decision contexts in science are rarely cut and dry. The implications of this point for the discovery/justification distinction, however, are less so. "What," we want to ask, "does the 'non-cut-and dryness' of decision contexts have to do with Reichenbach's distinction?" Kuhn answers as follows:

Considerations relevant to the context of discovery are then relevant to justification as well; scientists who share the concerns and sensibilities of the individual who discovers a new theory are *ipso facto* likely to appear disproportionately frequently among that theory's first supporters."[71]

Kuhn's point seems to be this: typical theoretical decision contexts are not straightforward or obvious; rather, each rival theoretical alternative will have reasons which support it. How scientists decide in such cases, then, depends on factors which are relevant to the discovery of the alternatives. In the oxygen/phlogiston example, factors relevant to the discovery of the oxygen theory (leaving aside for now the question of whether it makes sense to speak of theories as being "discovered") are likely to be factors relevant to the various members of the scientific community's decisions to adopt the oxygen theory: "scientists who share the concerns and sensibilities of the individual who discovers a new theory are *ipso facto* likely to appear disproportionately frequently among that theory's first supporters."[72]

This point may be right – it may be that factors relevant to the discovery of a theory are also relevant to other scientists' decisions to adopt that theory – but it has little bearing on the justification/discovery distinction. For, whatever factors led scientists to adopt the oxygen theory (for example), we still can ask questions about the justification of that decision. As noted already, the context of justification is distinct from the context of decision. So at most Kuhn has shown that factors relevant to the context of discovery are relevant to the context of decision. He has *not* shown what must be shown in order to cast doubt on Reichenbach's distinction – namely, that factors relevant to the context of discovery are relevant to the context of justification.

(c) Kuhn offers one more reason for doubting Reichenbach's distinction. This reason depends on Kuhn's claim that "subjective factors" are an ineliminable element of a scientist's decision to adopt a scientific theory. Much has been written concerning such "subjective factors," and we need not enter into a detailed consideration of them here.[73] For, even if Kuhn is right that such subjective factors are an ineliminable element of a scientist's decision to adopt some theory, it nevertheless remains the case that the factors relevant to such decisions are not, necessarily, relevant to the justification of such decisions. As before, Kuhn has confused the context of decision with the context of justification. And, as before, such confusion will not do. We certainly do want an account of the scientific decision-making context. But, we also want an account of other aspects of the scientific enterprise – in particular, of the justification of scientific decisions. Kuhn's reasons for denying Reichenbach's distinction all confuse decision with justification, as if deciding to adopt T is also to justify that decision. This however, is clearly a mistake – it is one thing to decide to adopt T, another to justify that decision. (If not, how could one ever make an unjustifiable decision?) Kuhn has at most shown that factors relevant to the context of discovery are relevant to the context of decision. Consequently, he has not provided us with good reason for rejecting Reichenbach's distinction. And so, he has not provided any reason to take less seriously his critics' criticisms of his conception of scientific theory choice.

Finally, Kuhn turns his attention to the notion of "subjectivity." His critics err, Kuhn claims, if they take his claim that theory choice necessarily involves subjective factors to mean that theory choice is "undiscussable" or "a matter of taste," and those critics who take him to be claiming the contrary "have seriously mistaken [his] position."[74] Here is the "soft" Kuhn, the Kuhn who seemingly denies the incommensurability thesis and its corollaries. Kuhn *must* be here denying the incommensurability thesis, because that thesis guarantees the "undiscussability" of rival paradigms. According to incommensurability, rival paradigms are incapable of being meaningfully neutrally compared; proponents of such paradigms inevitably talk past each other as they speak "across the revolutionary divide."[75] If theory choice between such incommensurable rivals is anything, it is undiscussable. So, if Kuhn denies that theory choice is undiscussable, he must (on pain of inconsistency) deny the incommensurability thesis. Yet, one page later, Kuhn writes as if the incommensurability thesis is alive and well, as he indicates that it is "difficult or, more likely, impossible" for "commensurable" theory choice to take place:

> . . . I simply assert the existence of significant limits to what the proponents of different theories can communicate to one another. The same limits make it difficult or, more likely, impossible for an individual to hold both theories in mind together and compare them point by point with each other and with nature. That sort of comparison is, however, the process on which the appropriateness of any word like "choice" depends.[76]

We have again, thus, the by now familiar tension which runs through Kuhn's later writings: the tendency to, on the one hand, defend a strong incommensurability claim; and, on the other, deny any strong incommensurability claim. As we have seen earlier, such tension is not acceptable – Kuhn simply cannot (consistently) have it both ways. This most recent defense, therefore, does little to rectify the problems with Kuhn's view of the objectivity or rationality of theory choice already discussed.

(3) *Relativism*

Finally, Kuhn's critics have also argued that Kuhn's incommensurability thesis renders his view relativistic, in that acceptance of a paradigm or theory is relative to one's prior commitment to the paradigm or theory in question, thus making objective evaluation impossible. In later responses to his critics, Kuhn has attempted to refine his original position concerning relativism, and to show that he is not a relativist in any "bad" sense:

> If you would think of science's developing in the usual evolutionary treelike pattern, then I think it not only possible, but indeed quite easy, to do the following: design a set of criteria which would permit a neutral observer to say which of two theories for the same aspect of nature was the later (the more evolved) and which was the earlier . . . In this sense scientific development is a unidirectional and irreversible process, and that is not a relativistic view.[77]

This passage, however, in no way saves Kuhn from relativism. Kuhn's earlier view was seen as relativistic because that view seemed to hold that acceptance of one theory over another had nothing to do with evaluation according to neutral, objective criteria of evaluation. Proponents of competing paradigms each judged their own paradigm superior to its rivals, according to criteria their own paradigm dictated. Consequently, a given paradigm (P_1) was held to be superior according to the criteria dictated by P_1, by the proponents of P_1; while a rival paradigm (P_2) was held to be superior according to the criteria dictated by P_2, by the proponents of P_2. This rendered paradigm-evaluation relativistic in the following sense: which paradigm was judged superior depended on which set of criteria of evaluation one judged by. P_1 was superior relative to the criteria dictated by P_1, but not relative to those dictated by P_2. P_2 was superior relative to the criteria dictated by P_2, but not relative to those dictated by P_1. Since Kuhn held that there were no neutral criteria available, evaluation was relative to the criteria, hence the paradigm, appealed to. This is the sense in which his view was seen as relativistic.

The evolutionary metaphor Kuhn suggests in the above cited passage does not speak to the relativism issue just outlined. For at best, the evolutionary criteria suggested by the metaphor will indicate which of a set of competing paradigms is later, or more developed. *But later does not mean better*. The proponents of the earlier paradigm (as picked out according to the criteria suggested by the evolutionary construal) could still judge their paradigm superior to the later one, according to the criteria of evaluation dictated by their (earlier) paradigm. Consequently, Kuhn's evolutionary metaphor does not in the least show his view to be non-relativistic. Here, as elsewhere, Kuhn's later remarks do little to make plausible, or to clarify, refine, or improve, his original presentation.

E. CONCLUSION

The three recent responses by Kuhn considered above are not exhaustive, but are, I believe, representative, of Kuhn's latest attempts to clarify and refine his position concerning incommensurability, relativism, and the objectivity of theory choice in science.[78] As the discussion thus far has shown, these further responses do little to render his view plausible. His position remains both confusing and confused.

Objectivity, as Scheffler has stated, "requires simply the possibility of intelligible debate over the comparative merits of rival paradigms."[79] If Kuhn allows for such debate, he must indeed retract his irrationality thesis, as well as his incommensurability thesis and so his relativism thesis. He claims to have done the first, while refusing to do the second. This, it seems clear, is untenable. If he indeed retracts the irrationality thesis, he capitulates to his critics by granting at least the possibility that theory choice can be objective. If he does not retract the irrationality thesis, his position is conceptually confused, and his introduction of "values" into the discussion only further confuses the issues. His claim that there is no algorithm for theory-choice likewise offers Kuhn little help, as it both fails to clarify his view, and is at any rate agreed to by all parties in the debate. Kuhn's more recent attempts to respond to his critics do not, finally, do so adequately. The irrationality thesis, the incommensurability thesis, and so Kuhnian relativism remain tenuous and unwarranted; Kuhn has failed to show that theory choice cannot be objective or rational, or that it must be relative to paradigm commitment. On the other hand, if Kuhn seriously agrees with his critics that science is not necessarily subjective or irrational, as he has claimed, he has failed to show how this sentiment can be maintained in conjunction with the rest of his views concerning the dynamics of theory change.

In this chapter we have been concerned only with Kuhn's own discussion of relativism, objectivity, rationality, and incommensurability. I hope to have shown that Kuhn's analysis of science is insufficient to warrant our embracing a relativistic view of scientific knowledge. That is, Kuhn has not offered even a minimally compelling argument for relativism (or incommensurability, etc.). However, it is not sufficient to consider only Kuhn's writings on this score, for several other writers have offered defenses and extensions of Kuhn's analysis, including his arguments for relativism. In the next two chapters we will examine the most important of these defenses.

CHAPTER 4

THE KUHNIANS

Several philosophers have come to Kuhn's aid in the debate concerning objectivity, rationality, incommensurability, and relativism. In this chapter I examine two of these defenses/extensions of the Kuhnian position. The first centers on the objectivity of science, in which Jack W. Meiland seeks to interpret Kuhn as a proponent of objectivity, contrary to the standard critical interpretation of Kuhn as discussed in Chapter 3. The second is a reinterpretation and defense of Kuhnian relativism by Gerald Doppelt. Doppelt's focus on relativism makes his work particularly relevant to this volume. (In the next chapter I consider the general conception of the "new philosophy of science" built along Kuhnian lines offered by Harold I. Brown. Brown's discussion also concerns itself, in part, with relativism.) In the case of each of these authors I will try to show that their discussions fail to support a Kuhnian conception of science in general, or of relativism in particular. I begin with Meiland.

(A) MEILAND ON KUHN AND OBJECTIVITY

As we saw in Chapter 3, several of Kuhn's critics, including Scheffler, Shapere, and Kordig, have interpreted Kuhn as offering an account of scientific knowledge in which that knowledge is non-objective. At the least, according to this interpretation, Kuhn's conception of scientific theory choice is non-objective, in that incommensurability precludes objective evaluation of rival paradigms. (And of course this interpretation has it that this subjectivity is closely related to relativism, since theory choice is relative to prior paradigm-commitment. Thus our present discussion, while couched in terms of objectivity, is highly relevant to our larger theme of relativism.) Meiland, in his article "Kuhn, Scheffler, and Objectivity in Science,"[1] challenges this interpretation of Kuhn, and argues that, these critics notwithstanding, Kuhn does not portray science as subjective (or relative or irrational). In this section of the present chapter we examine Meiland's analysis of the Kuhn/Scheffler debate regarding the objectivity of science.

Meiland argues that Kuhn has been wrongly interpreted "as an opponent of objectivity when in fact he is not an opponent but a proponent."[2] Meiland also argues that this point, if accepted, will "shift the focus [of the debate between Scheffler and Kuhn] from objectivity to the real issue, namely truth."[3] I propose to take these claims up in order.

(1) *Is Kuhn an Opponent or a Proponent of Objectivity?*

Scheffler identifies Kuhn's main thesis as the thesis that "paradigm change in science is not generally subject to deliberation and assessment."[4] Meiland, however, argues that Kuhn is a proponent of objectivity not with respect to theory (paradigm) change, but rather with respect to two other aspects of science: what Meiland calls "IS objectivity" and "SIS objectivity." Let us turn to these briefly.

IS objectivity can be characterized by the existence of *independent standards* (hence "IS") which can serve as tests of a theory (paradigm). Meiland argues that Kuhn is a proponent of IS objectivity, at least insofar as observation is concerned, for Kuhn argues that anomalies are the things which provoke the crises which set the stage for a scientific revolution:

> Discovery commences with the awareness of anomaly, i.e., with the recognition that nature has somehow violated the paradigm-induced expectations that govern normal science. It then continues with a more or less extended exploration of the area of anomaly. And it closes only when the paradigm theory has been adjusted so that the anomalous has become the expected.[5]

It is certainly true that Kuhn depends on the notion of anomaly in explaining scientific revolution. However, given his explication of anomaly (above) and his characterization of paradigms, it is not clear that he is entitled to do so. For Kuhn argues that communication across paradigms is circular and at cross-purposes, and more fundamentally, that *competing paradigms are incommensurable*. If the incommensurability claim is taken seriously, however, there can be no paradigm-neutral "nature" capable of violating "paradigm-induced expectations." The problem is this: Kuhn cannot have both anomaly and incommensurability. For the former notion assumes a paradigm-neutral nature, but such a paradigm-neutral standard eliminates the possibility of incommensurability of competing paradigms. For incommensurability is simply the absence of such a paradigm-neutral standard.

Meiland recognizes this problem, and tries to solve it by "holding firm" the principle "that there can be clashes between paradigms and nature."[6] He argues that if we allow Kuhn anomalies, we can keep his system consistent by confining irrationality to choice among incommensurable paradigms. Unfortunately, this line does not work. For if we "hold firm" the notion of anomaly, we must give up the incommensurability of competing paradigms, for we will have a paradigm-neutral standard – namely, nature. Contrary to Meiland, it makes little difference "which end of Kuhn's views we start from."[7] If we start with anomaly, we must give up incommensurability and irrational choice; if we start with incommensurability, we must give up any notion of a paradigm-neutral "nature" and anomaly. From either end, the conjunction of anomaly and incommensurability is inconsistent.

Meiland's second brand of objectivity, "SIS objectivity," fares no better than the first. SIS objectivity differs from IS objectivity in that SIS objectivity demands that competing theories have the *same* independent standards, not just independent standards, each theory with its own. Meiland claims that Scheffler claims that Kuhn's theory does not allow for SIS objectivity; and consequently there is, for Kuhn, "no way to judge scientific progress."[8] Meiland then employs a two-stage strategy to defend Kuhn: (i) Meiland tries to revise Kuhn so that, even if Kuhn does not allow SIS objectivity, he can still account for scientific progress; (ii) Meiland tries to expand the notion of SIS objectivity to include non-observational shared independent standards, so that Kuhn can have such an "enlarged" SIS objectivity in his theory, thus again qualifying as a proponent of objectivity. Let us look at these moves in order.

(i) Can Kuhn account for progress in science? Kuhn tries to account for it by claiming that "the new paradigm should not lose previously-acquired problem solving ability and should seem, moreover, to solve new problems not otherwise soluble."[9] So progress can be measured in terms of *problem-solving ability*. Kuhn is immediately confronted with a problem, pointed out by Meiland: "If theories are conceptually incommensurable in the way in which Kuhn describes in his book, can the puzzles generated by the old paradigm even be formulated within the conceptual structure (or way of seeing the world) of the new paradigm?"[10] Meiland argues that "there is . . . no theoretical bar in Kuhn's position to there being *some* dimension along which theories can be positioned and thus evaluated."[11] This is simply false. There *is* such a theoretical bar – namely, incommensurability. If such a dimension were permissible for Kuhn, there would be no incommensurability between paradigms, and no irrationality in paradigm choice. Even if Meiland were right on this point, his suggestion – that a paradigm's problem-solving ability be measured according to how many problems that *it* generates, it solves – would in all likelihood never allow us to choose a new theory.[12] For, even when a paradigm is in crisis, it will have amassed a great amount of problem-solving ability. This is to be expected, for periods of normal science – that is, work done under a single paradigm – are devoted to amassing solutions to the problems, or puzzles, the paradigm suggests. Particularly in the case of a relatively old, well-entrenched paradigm, it is unrealistic to expect its potential successor to show itself superior in terms of problem-solving ability. Kuhn himself recognizes this point: "The man who embraces a new paradigm at an early stage must often do so in defiance of the evidence provided by problem-solving."[13]

So, even if there were no "theoretical bar in Kuhn's position to there being *some* dimension along which theories can be positioned and thus evaluated,"[14] it seems impossible to propose such a dimension which does not beg the question by being paradigm-bound, and which also allows, realistically, for paradigm change. And of course, as mentioned above, there *is* such a "theoretical bar" – namely, the incommensurability of paradigms. The problem can be put as a dilemma: either this evaluative dimension is paradigm-bound, or it is not. If the former, it will not work; if the latter, Kuhn is not entitled to it (unless he is willing to revoke the claim of incommensurability).

(ii) Let us now look to Meiland's second argument, which purports to show that, sufficiently broadened, Kuhn could be said to allow SIS objectivity within his theory. Meiland's suggestion is simply that we broaden the notion of SIS objectivity so that *non-observational* criteria are admitted for candidacy as standards. Then, Kuhn could claim SIS objectivity according to some non-observational standard which would not be affected by Kuhn's claim regarding the incommensurability of competing paradigms. As argued above, such a move is futile. If we take the notion of incommensurability seriously, there can be no such paradigm-neutral standard – observational or otherwise – with which to rank competing paradigms.

When discussing the problem of these standards being paradigm-bound, furthermore, Meiland commits a crucial blunder. He incorporates into Kuhn's position a view of Scheffler that, if Kuhn accepted, would end the disagreement between them – namely, Scheffler's view that first-order paradigm differences do not necessarily lead to second-order differences, that is, differences in the paradigms which are applied to nature itself do not entail differences in the evaluative criteria which are to be applied in judging those paradigms as satisfactory or not.[15]

This is one of the crucial points Scheffler makes against Kuhn and is, contrary to Meiland, not at all consistent with the radical Kuhnian position discussed in the last chapter. If this were Kuhn's view, there would be no reason to suppose competing paradigms incommensurable, no reason for paradigm debates to be circular, no reason for most of Kuhn's more radical views. In any case, Kuhn holds not this view of Scheffler's that first-order differences need not lead to second-order differences, but rather, as Scheffler describes it, the view that "two scientists of differing first-order persuasions [will] inevitably step back to different evaluative positions."[16] Witness the following passage from Kuhn:

In learning a paradigm the scientist acquires theory, methods, and standards together, usually in an inextricable mixture. Therefore, when paradigms change, there are usually significant shifts in the criteria determining the legitimacy both of problems and of proposed solutions.

That observation . . . provides our first explicit indication of why the choice between competing paradigms regularly raises questions that cannot be resolved by the criteria of normal science . . . . In the partially circular arguments that regularly result, each paradigm will be shown to satisfy more or less the criteria that it dictates for itself and to fall short of a few of those dictated by its opponent.[17]

It is hard to read this passage in any other way than that second-order differences *do* result from first-order ones. Scheffler, in commenting on the passage, argues that it suggests that:

any sharing of second-order standards is impossible. For to accept a paradigm is to accept not only theory and methods, but also governing standards or criteria which serve to justify the paradigm as against its rivals, in the eyes of its proponents. Paradigm differences are thus inevitably reflected upward, in criterial differences at the second level.[18]

Meiland's importation of Scheffler's views to Kuhn's cannot but obscure matters. There is no denying that there *is* a substantial disagreement here – Kuhn is claiming that paradigm debate is not objective, but rather, that it is irrational, circular, subject to paradigm-bound criteria, etc.; while Scheffler is claiming that paradigm debate is not necessarily so. Here is the core of the dispute between Kuhn and Scheffler: Can paradigm debate be objective? Kuhn says no, Scheffler says yes.

Meiland admits that it is Kuhn's view that the adoption of paradigms is irrational;[19] it seems odd, then, that Meiland holds that Kuhn "is not an opponent of objectivity but a proponent."[20] Presumably, "proponent of objectivity" means "proponent of objectivity across the board," not "proponent most of the time, but opponent in case of paradigm debate or choice." In any case, Meiland's claim that Kuhn is a proponent of objectivity must clearly exclude the area of paradigm debate; and it is precisely in this area that Scheffler takes issue with Kuhn. Meiland has erected, for the purposes of Scheffler's discussion of Kuhn, straw issues in the form of IS and SIS objectivity. It is whether or not theory change can be objective that is the concern of Scheffler in his debate with Kuhn. And at least as far as the issue of change is concerned, Meiland admits[21] that Kuhn's view is not pro-objectivity. So, we must answer Meiland's first claim, namely, that Kuhn has been wrongly interpreted "as an opponent of objectivity when in fact he is not an opponent but a proponent"[22] in this way: even if Kuhn were a proponent of IS and SIS objectivity (I have argued above that he cannot be consistently so interpreted), he is an opponent of objectivity as regards theory change – and it is with this issue that Scheffler is most concerned.

(2) *Is the "Real Issue" between Scheffler and Kuhn Truth?*

Meiland argues, in the final portion of his paper, that there really is no disagreement between Scheffler and Kuhn as far as objectivity goes; that the "real" underlying issue between Kuhn and Scheffler concerns the notion of truth – specifically, Kuhn's denial that there is a sense in which adopted theories are *more true* (better reflect reality) than their predecessors:

> Nevertheless, there is another step, or kind of step, which many philosophers of science wish to take and which I refuse. They wish, that is, to compare theories as representations of nature, as statements about 'what is really out there'. Granting that neither theory of a historical pair is true, they nonetheless seek a sense in which the latter is a better approximation to the truth. I believe nothing of that sort can be found.[23]

Now it is not clear whether this point concerns the use of truth as a *criterion* of theory selection or, also, all comparisons of truth of theories selected by *independent criteria*.[24] Scheffler goes to some trouble to spell out his position that, while truth is not a *criterion* of theory choice (that is, we do not pick one theory over another *because* one is "more true"), the theory chosen on independent grounds of credibility has a stronger claim to represent the truth than a theory rejected on those grounds.[25] If there is a disagreement here between Scheffler and Kuhn, is it in any case the crucial disagreement? I think not. In trying to understand Scheffler's differences with Kuhn, we need only to look to Scheffler's arguments. Here, he is crystal clear: "The issue is . . . *objectivity*, and objectivity requires simply the possibility of intelligible debate over the comparative merits of rival paradigms."[26] So, on Scheffler's own word, his objection to Kuhn concerns objectivity, not truth; and furthermore, objectivity with respect to paradigm debate – which, Meiland acknowledges, is subjective on Kuhn's account.[27]

(3) *General Problems with Meiland's Account*

(1) Meiland wants to argue that Kuhn is not an opponent of objectivity, but rather a proponent; and to that end he cites Scheffler to support Kuhn (discussed above, Section 1) that "first-order paradigm differences do not necessarily lead to second-order differences."[28] If Kuhn holds this view (I have argued above that he does not), why should Meiland argue that, on Kuhn's view, "irrationality in science is confined to the adoption of paradigms"?[29] If Kuhn believes that first-order diffferences do not lead to second-order differences, why should he hold irrationality anywhere? For the irrationality of paradigm choice is due solely to the second-order criterial differences resulting from the first-order differences. If the second-order differences go, so do the irrational choices.

(2) Meiland claims both that Kuhn is a proponent of objectivity,[30] and also that, on Kuhn's view, paradigm choice is irrational.[31] How can these views both be maintained? If paradigm choice is not an objective affair, and Kuhn holds that it is not, then Kuhn can be called a proponent of objectivity in science only if paradigm choice is not a part of science. But that is surely not Kuhn's intent – his concern is to explain that process, called "scientific revolution," which is the cornerstone of scientific endeavor. Indeed, it would be hard to find one more emphatic about the importance of paradigm shift in science than Kuhn. Now, if it is granted that paradigm choice is part and parcel of science, then I see no way for Meiland to reconcile his two claims, that, on the one hand, Kuhn is a proponent of objectivity in science, and that, on the other, Kuhn holds paradigm debate to be non-objective. It is clear that Meiland cannot have it both ways.

Meiland's interpretation/defense of Kuhn, then, must be judged unsuccessful. Neither of Meiland's central contentions withstand analysis: it is not the case that Kuhn is a proponent of objectivity; nor that the debate between Scheffler and Kuhn is really over truth, rather than objectivity. On the contrary, it is clear that 1) Kuhn cannot be construed as a proponent of objectivity in science, especially insofar as theory change is part of science; and 2) that the "real" quarrel between Scheffler and Kuhn is not over truth, but over the objectivity of paradigm debate. Finally I have tried to show 3) that Meiland's article is inconsistent in that it holds both that Kuhn is a proponent of objectivity in science in general, but a proponent of non-objectivity in one major aspect of science, namely theory choice. This seems a flat-out contradiction, and can only be remedied by Meiland acknowledging that while Kuhn may be a proponent of objectivity in *some* areas of science, he is not so with respect to theory choice. In this case, his contention that the issue between Kuhn and Scheffler is not objectivity, but truth, must fall, for the issue will be, by his own admission, over objectivity and theory change – which has been Scheffler's claim from the start.

(B) DOPPELT ON KUHN AND RELATIVISM

While Meiland, as we have seen, tries to soften the Kuhnian position so that it yields a sort of objectivity, Gerald Doppelt offers an interpretation of Kuhn which firmly accepts, and indeed emphasizes, a straightforwardly relativistic conception of scientific knowledge. Doppelt's paper[32] offers and defends a reconstructed version of Kuhn's "incommensurability thesis" and, relatedly, Kuhn's epistemological relativism. Here I will argue that Doppelt's reconstruction fails, and that the incommensurability thesis and Kuhnian relativism remain unwarranted. I begin by reviewing Doppelt's reconstruction.

(1) *Doppelt's Reconstruction of Kuhn*

The central tenet of Doppelt's reconstruction of Kuhn is that "it is the incommensurability of scientific *problems* between rival paradigms and not that of *meanings*"[33] which underlies Kuhnian relativism. According to Doppelt, incommensurable paradigms are *not* imprisoned in their own language. Rather, they set for themselves different and incompatible *criteria of explanatory adequacy*. These incompatible standards are generated from each paradigm's tendency to disagree as to what counts as the fundamental problems any paradigm in the field ought to solve. So, consider rival paradigms P_1 and P_2. Each paradigm has its own set of fundamental problems which it deems to be *the* basic set of problems that *any* paradigm in the discipline must satisfactorily answer. Let 'Pr' stand for problems, so that P_1 recognizes Pr_1 as the set of fundamental problems any paradigm in the discipline must satisfactorily answer, while P_2 recognizes Pr_2 as that privileged set. Since P_1 and P_2 typically recognize incompatible sets of problems as basic, each paradigm maintains incompatible standards, or criteria, of explanatory adequacy. Each paradigm's standards are based on that paradigm's set of fundamental problems, so that P_1 has standards of adequacy S_1, based on the set of problems P_1 recognizes as fundamental – namely, Pr_1. Analogously, P_2 has standards of adequacy S_2 based on *its* set of fundamental problems, Pr_2. Since the sets of problems each paradigm recognizes as fundamental are disparate, so also are the standards of adequate solution. Incommensurability lies, then, in the differing standards of explanatory adequacy each paradigm adopts, both for itself and for the discipline in general:

> [W]hat makes the different problems of concern to rival paradigms "incommensurable" is not their mere difference, but the fact that these differences are incorporated into the standards of explanatory adequacy each paradigm implicitly sets for itself and all theory.[34]

Doppelt uses Kuhn's example of the shift from pre-Daltonian to Daltonian chemistry to illustrate the incommensurability of standards. Pre-Daltonian chemistry, Kuhn and Doppelt argue, had achieved a substantial amount of success in answering certain questions, and solving certain problems, which that paradigm took to be basic. For example:

> The old chemistry was able to explain the observable qualities of chemical substances – e.g. why the metals were so much more alike in their observed metalline qualities than their ores, and also the qualitative changes they undergo during chemical reactions, such as the formation of observed acidic properties.[35]

The new, Daltonian chemistry did not, argue Kuhn and Doppelt, simply adopt the explanatory power of the old chemistry and recognize the importance of the questions and problems the old chemistry took to be central to chemical inquiry. Rather, the new chemistry by and large "abandoned any concern for these questions and these observational data"[36] which the old chemistry took to be central, and substituted as the central concern of chemical theory different problems – "quantitative" problems "concerning weight relations and proportions in chemical reactions"[37] – which the old chemistry did not take to be central. What happened in the Daltonian chemical revolution was, then, a shift in the problems thought to be central to chemical inquiry, and so a shift in the standards of explanatory adequacy of chemical theory – for each paradigm judged chemical theory according to how well such theory succeeded in explaining the problems *it* took to be central to chemical inquiry. The Daltonian revolution should not be viewed simply as the expansion of chemical theory to include the phenomena which the old chemistry accounted for, as well as the phenomena which the new chemistry accounted for, according to Doppelt and Kuhn, because the new chemistry in fact *lost* much of the ability to account for the phenomena the old chemistry could account for. Rather, the new chemistry ceased to be fundamentally interested in the problems the old chemistry took to be basic; the new chemistry instead relegated those problems to the back burner (or indeed, declared them "unscientific"), and took as central to chemical inquiry a new set of problems ("quantitative" aspects of chemical reactions) regarded as less than crucial by the old chemistry. As Doppelt summarizes the Dalton example:

> Dalton's paradigm *cannot* be characterized as simply offering a better and more rational account of chemical phenomena than its predecessor; for the two paradigms seek to explain different kinds of observational data, in response to different agendas of problems, and in accordance with different standards of success.[38]

It is important to note that on Doppelt's interpretation of Kuhn, incommensurable paradigms *can* share certain concerns and "empirical points of contact" – in the Dalton case, for example, both pre-Daltonian and Daltonian chemistry recognized as a problem for chemical theory the explanation of the phenomenon of weight-gain in combustion.[39] Incommensurability thus does *not* lie in a total conceptual separation between competing paradigms. Rather, on Doppelt's reconstruction, incommensurability lies in the different relative importance rival paradigms assign to various problems and data, which insures differences in standards:

> [T]he very substance of . . . differences in data and observational concerns is incorporated into the standards of good theory implicit in the rival paradigms.[40]

The incommensurability of concepts, taken as basic to Kuhn's view by some of Kuhn's critics (e.g. Scheffler and Shapere),[41] Doppelt takes to be not central to his reconstruction of Kuhn. Rather, Doppelt takes the incommensurability of concepts to be derivative from the more basic incommensurability of rival paradigm's "conceptions of science":

[T]he root sense in which the concepts of rival paradigms are incommensurable is that they are embedded in incommensurable conceptions of science (involving incommensurable data, problems, and standards).[42]

To sum up: Doppelt's reconstruction of Kuhn allows for a fair amount of "logical contact" between rival paradigms. Nevertheless, such rivals are, on Doppelt's account, incommensurable in that they embody incompatible attitudes toward the fundamental problems any paradigm in the field ought to (try to) solve. Since paradigms disagree as to what the fundamental questions are, they disagree as to the proper standards of explanatory adequacy by which any paradigm in the field must be assessed, because each paradigm's standards will be a function of the set of problems each paradigm recognizes as fundamental to the discipline. Epistemological relativism, on Doppelt's account, results from the incommensurability of standards of explanatory adequacy of rival paradigms. Since paradigms are incommensurable in this respect, a paradigm's evaluation is relative to the standards of adequacy of the paradigm from which one is evaluating. A paradigm will be assessed variously according to how well it meets the standards of adequacy of various paradigms – thus, P_1 will be (typically) superior to P_2, relative to the standards of adequacy of P_1, while P_2 will be superior to P_1, relative to the standards of adequacy of P_2. Since P_1 and P_2 are incommensurable, they do not share common criteria of adequacy (though they may well share certain items of observational data, problems, and concepts); and since their criteria of adequacy are incompatible (if not, P_1 and P_2 would not be incommensurable), their assessment is relative to the paradigm-bound criteria of adequacy appealed to in making such an assessment.

Doppelt's reconstruction of Kuhn raises several questions:

(1) The revised conception of incommensurability Doppelt espouses is considerably weaker than Kuhn's original treatment (at least in so far as Scheffler and Shapere construe it). Does this weakening of incommensurability render it trivial? That is, is there anything *interesting* about incommensurability, and epistemological relativism, as depicted by Doppelt? What is the "cash value" of incommensurability, and relativism, on Doppelt's reconstruction of Kuhn?

(2) Doppelt's reconstruction hinges on the incommensurability of competing standards of adequacy of rival paradigms. Such incommensurability depends on the absence of paradigm-neutral *external* standards of adequacy by which a paradigm's internal standards can be non-relativistically evaluated. Does Doppelt's reconstruction of Kuhn adequately account for this internal/external distinction?

(3) Doppelt defends a version of epistemological relativism. Yet, as I argued in Chapter 1, epistemological relativism generally, and a fortiori Doppelt's version, is self-refuting. Can Doppelt defend his version of epistemological relativism from the *reductio* charge?

These questions will be taken up in the following sections.

(2) *Doppeltian Incommensurability*

The revised version of incommensurability which Doppelt defends denies that there is an "absolute epistemological break between mutually 'incommensurable' paradigms which *a priori* imprison themselves within their respective language, data, problems, and standards of science, and blocks the possibility of any coherent debate or comparison between them."[43] Rather, incommensurability amounts to the claim that:

> [T]here are basic, non-cumulative differences in the observational data and problems of concern to rival paradigms which are incorporated into their respective standards of theoretical adequacy.[44]

This rendering of incommensurability is substantially weaker than Kuhn's original treatment (as interpreted by many of Kuhn's critics, at any rate) in that it explicitly denies that incommensurability "blocks the possibility of any coherent debate or comparison"[45] between rival paradigms. That is, on Doppelt's reconstruction of Kuhn, P_1 and P_2 may be incommensurable, yet proponents of these rival paradigms can meaningfully and coherently debate the merits of their respective paradigms.[46] If the possibility of such debate is granted, one wonders about the importance of Doppelt's revision of incommensurability. What is the power or importance of incommensurability if rational debate between proponents of rival paradigms is possible?

Consider the Daltonian chemical revolution. Doppelt argues that Daltonian and pre-Daltonian chemistry are incommensurable in that they have fundamentally different conceptions of chemical science –

they disagree as to the membership of the set of fundamental problems of chemistry, the adequacy of purported solutions to these problems, and their relative importance. Moreover, since each paradigm's standards of theoretical adequacy are a function of these very considerations, each paradigm will evaluate all chemical paradigms according to standards of adequacy dictated by that paradigm's fundamental conception of chemical science. Hence, it is highly improbable (if not in principle impossible) that judgments of the relative merits of rival paradigms will ever reach consensus – each paradigm will, rather, favor itself in comparison with its rivals. This is, seemingly, a strong incommensurability. However, Doppelt's admission that coherent debate and comparison of rival paradigms is possible vitiates any strong incommensurability position. For if such debate and comparison are possible, why should it be that rival paradigms inevitably judge themselves superior? Why, given the substantial "logical contact" between rival paradigms Doppelt allows, is rational consensus doomed? Given meaningful debate and comparison "across the revolutionary divide," what sort of epistemological relativism can there be?

Doppelt's answer is straightforward: meaningful debate and comparison are not sufficient to insure rational consensus, because rival paradigms

> lack a sufficiently common definition of the discipline and its criteria of explanatory adequacy to allow their discourse to terminate in rational consensus.[47]

This lack of common definition, Doppelt argues, results in "fundamental disagreements, irresolvable by scientific argument."[48] However, there is no reason to believe that such debate is necessarily irresolvable. Once we are granted the possibility of coherent debate and comparison, there is no reason to believe that such debate will necessarily fail to result in consensus. There is no doubt that the proponents of pre-Daltonian and Daltonian chemistry adhered to radically different views concerning the fundamental subject-matter and problems of chemistry. Nevertheless, given the possibility of coherent debate, there is no reason to believe that these radical differences are irresolvable. Once the possibility of coherent debate is granted, we are simply overly pessimistic to think that radical differences can *never* be overcome. After all, Dalton had *good reason* to think that "quantitative problems concerning weight relations and proportions" were deeper, and more basic to chemical theory, than the qualitative questions addressed by pre-Daltonian chemistry. His decision to place his new set of fundamental problems at the center of chemical theory was not arbitrary.[49] Given the possibility of coherent debate between Dalton and pre-Daltonians, then, it is possible for pre-Daltonians to seriously consider, and fairly evaluate, Dalton's new picture of chemical inquiry. And given *that* possibility, it is misleading to declare such differences – which are amenable to rational debate and coherent comparison – incommensurable, for incommensurability signifies precisely the in-principle impossibility of such debate and comparison.

Doppelt, presumably, would reply that the mere possibility of coherent debate between proponents of rival paradigms is insufficient to refute incommensurability, since, while rational debate and coherent comparison are possible, proponents of rival paradigms will nevertheless fail to reach rational consensus, because their standards of theoretical and explanatory adequacy are sufficiently disparate to block consensus. This reply is ineffectual, however, because these different standards of adequacy are themselves, on Doppelt's account, open to meaningful comparison and rational evaluation. Given that standards of adequacy are themselves open to coherent comparison and debate, there is no reason to assume that rational consensus is impossible. Witness again the previously cited passage:

> [T]his thesis [the incommensurability thesis] does *not* imply an absolute epistemological break between mutually "incommensurable" paradigms which *a priori* imprison themselves within their respective language, data, problems, and *standards of science*, and blocks the possibility of any coherent debate or comparison between them.[50]

This passage clearly allows that standards of science, as well as language, problems, and data, can be meaningfully compared and coherently debated. And this gives away the game. The admission of the possibility of rational debate concerning the standards of adequacy of rival paradigms vitiates any powerful incommensurability or relativistic position.

Doppelt attempts to limit the coherent debate concerning the merits of rival paradigms by distinguishing between good reasons and compelling reasons for paradigm choice. He argues that the proponent of a paradigm can recognize the existence and force of reasons which favor its rival (for example, a pre-Daltonian chemist "can recognize that the empirical success of the new chemistry in explaining [the old chemistry's] anomalous cases of weight-gain in combustion is a (good) reason on behalf of [the new chemistry]"[51]), but that such good reasons are never compelling: "The recognizable good reasons which the advocates of a new paradigm urge upon the practitioners of the established paradigm are never 'compelling' as long as the latter adhere to the standards of science implicit in their paradigm and its achievement."[52] By drawing a distinction between good reasons and compelling reasons, Doppelt seeks to preserve a rational dimension of theory choice, while at the same time maintaining incommensurability. There is, however, a problem with his distinction between good and compelling reasons. Doppelt draws the distinction between good and compelling reasons according to their ability to undermine a scientist's adherence to the standards of science implicit in her paradigm. A reason can never be compelling, according to Doppelt, as long as the scientist adheres to her paradigm's standards of science. However, as was argued earlier, such adherence is itself not blind or uncriticizable; alternative standards of science can themselves be meaningfully compared and coherently debated. Therefore, reasons can be brought to bear on the decision whether or not to adhere to a given paradigm's standards of science. There is no "reason-proof" shield which protects one's adherence to a paradigm's standards of science. And so, there is no reason to think that there can never be reasons for paradigm-change which are "compelling." Doppelt's distinction between good and compelling reasons, and his claim that reasons for paradigm change can never be compelling, are undermined by his admission that standards of science are themselves open to meaningful comparison and coherent debate.

Doppelt's reconstruction is thus doomed. Doppelt wants to defend an interesting, powerful version of incommensurability and epistemological relativism. He also wants to allow a "considerable overlap between the language, problems, data and standards of rival paradigms"; an overlap sufficient to allow and account for a "shared context of rational debate"[53] across paradigms. But these, I hope to have shown, are incompatible, and their conjunction inconsistent.

The above analysis hinges on the possibility of rational debate by proponents of rival paradigms concerning their respective paradigm's criteria of theoretical adequacy. Such debate must, presumably, depend on the possibility of paradigm-neutral perspective: that is, such debate depends on the existence of paradigm-neutral meta-standards by which paradigm-bound standards can be paradigm-neutrally evaluated. This distinction between paradigm-neutral external standards and non-neutral internal standards is a crucial feature of the Kuhn/Doppelt-Scheffler/Shapere debate, and a close look at the distinction with respect to Doppelt's reconstruction of Kuhn's incommensurability and relativism theses is warranted. To this we now turn.

(3) *Internal and External Standards*

I have argued thus far that the possibility of rational debate and coherent comparison of rival paradigm's standards of adequacy (which Doppelt is committed to on his reconstruction of Kuhn) renders incommensurability an empty and trivial view. The reason for this reduction in the power of the incommensurability thesis is obvious: since rational debate and coherent comparison are possible, there must be paradigm-neutral vantage points from which rival paradigms may be meaningfully compared (for if not, then rational debate and coherent comparison would be impossible). If rational debate is possible, then, while internal standards of adequacy may be operative within a paradigm, there must be external standards by which internal standards can themselves be neutrally judged.

Doppelt, as we saw earlier, wants to maintain both an interesting and powerful incommensurability thesis, and also sufficient cross-paradigm connections to insure a "shared context of rational debate."[54] In the context of the internal/external distinction, this amounts to Doppelt's defending a muddled version of that distinction. Doppelt argues that there is such a distinction, but that external standards are determined by internal standards, and that "a relativism of internal standards implies a relativism of external standards as well."[55] Why this account of the internal/external distinction is unsatisfactory is worth spelling out here.

Doppelt, as the above citation indicates, holds that there are external standards, but that they are determined by internal ones. (How an external standard could be determined by internal standards and yet still be external is not at all clear, but we shall let this problem pass.) He also holds[56] that there are no operative external standards, and that the strength of reasons in paradigm debate is "relative to the disparate internal standards of rival paradigms."[57] This inconsistency may be excused as largely a verbal one; it is clear that his defense of incommensurability demands the absence of paradigm-neutral external standards capable of non-arbitrarily resolving paradigm disputes. However, as we have seen several times already, Doppelt's allowance of rational debate and meaningful comparison commits him to some paradigm-neutral external standards. As it stands, then, his position is inconsistent, as it necessitates both the existence of, and the absence of, neutral external standards of adequacy. If there are, on Doppelt's view, paradigm-neutral external standards, then there can be no (interesting) incommensurability. If, on the other hand, there are no such standards, Doppelt cannot claim (as he has) that meaningful debate and comparison are possible. This (the absence of neutral standards) would make it necessary for Doppelt to reinstate the strong, "total conceptual imprisonment" version of incommensurability, which is open to the Scheffler-Shapere criticism discussed and developed in the previous chapter and which Doppelt has tried so hard to avoid.

(4) *Relativism and the Problem of Self-Refutation*

A final problem facing Doppelt's reconstruction of Kuhn is a problem facing any version of epistemological relativism – namely, the problem of avoiding self-refutation. As I argued in Chapter 1, epistemological relativism generally is open to the self-refutation charge, and some (including Scheffler, Shapere, and myself in Chapter 3) have argued that Kuhn's epistemological relativism, in particular, suffers from this fundamental flaw. Can Doppelt defend his version of epistemological relativism from self-refutation?

In order to set the stage for Doppelt's defense against the self-refutation charge, let us briefly review the sense in which Kuhn is a relativist. As noted in Chapter 3, Kuhn's view is relativistic in that it holds that the acceptance of one theory over another has nothing to do with evaluation according to paradigm-neutral, objective criteria of evaluation. Proponents of competing paradigms each judge their own paradigm superior to its rivals, according to criteria their own paradigm dictates. Consequently, a given paradigm (P_1) is held to be superior according to the criteria dictated by P_1, by the proponents of P_1; while a rival paradigm (P_2) is held to be superior according to the criteria dictated by P_2, by the proponents of P_2. This renders paradigm-evaluation relativistic in the following sense: which paradigm is judged superior depends on which set of criteria of evaluation one judges by. P_1 is superior relative to the criteria dictated by P_1, but not relative to those dictated by P_2. P_2 is superior relative to the criteria dictated by P_2, but not relative to those dictated by P_1. Since Kuhn argues that there are no neutral criteria available, evaluation is therefore relative to the criteria, hence the paradigm, appealed to. This is the sense in which his view is relativistic. It is well captured, I think, by the formulation of epistemological relativism (ER) offered in Chapter 1.

Doppelt's reconstruction differs from Kuhn's original position slightly. For one thing, Doppelt shifts the basis of incommensurability (and so, a key premise in the argument for relativism) from the incommensurability of concepts to the incommensurability of problems, data, and standards of adequacy. For another, Doppelt allows, as we have seen, for the existence of paradigm-neutral standards of evaluation. However, as I argued in the previous section, Doppelt's allowance of paradigm-neutral external standards is vitiated by his view that such external standards are determined by paradigm-bound internal standards: "a relativism of internal standards implies a relativism of external standards as well."[58] Thus, while Doppelt allows for the existence of external standards, they are not paradigm-neutral standards, for they cannot, according to Doppelt, be neutrally used to evaluate rival paradigms. Since such neutral application is necessary for a non-relativist position, it is clear that Doppelt's reconstruction is as relativist as Kuhn's original formulation. It too is captured by the formulation of epistemological relativism (ER) offered in Chapter 1. Consequently, Doppelt's reconstruction is equally open to the self-refutation charge.

If the arguments against ER developed in Chapter 1 are sound, then Doppelt's reconstruction of Kuhnian epistemological relativism succumbs to them, as does any epistemological relativist position. Doppelt seeks to establish a relativist position; but, in so far as he succeeds, the self-refutation arguments undermine that success, for the very principles and arguments Doppelt relies on in establishing his position are undermined by the self-refutation arguments. Doppelt claims that "our whole presentation of Kuhn's relativism disputes the assumption that any such relativism is a 'reductio'."[59] However, Doppelt's whole presentation does no such thing. The self-refutation charge is not a blind assumption; it consists in specific arguments demanding specific attention. Doppelt offers no specific rebuttal to any form of the self-refutation arguments; it is therefore appropriate to conclude that he has not circumvented the difficulties the self-refutation charge presents.

(5) *Preliminary Conclusion*

While Doppelt's reconstruction of Kuhn shifts the basis of incommensurability from concepts to problems, data, and standards, thereby allowing for a certain amount of "logical and empirical contact" between rival paradigms, it does not overcome the difficulties enumerated above. Specifically, Doppelt's reconstruction fails to defend an interesting version of incommensurability; fails adequately to treat the internal/external distinction so crucial to his (and Kuhn's) argument; and fails to come to grips with the self-refutation arguments against epistemological relativism. Thus I conclude that Doppelt's reconstruction of Kuhn does not present a defensible version of epistemological relativism.

(6) *Further Discussion of Doppelt's "Moderate" Relativism*

In his "A Reply to Siegel on Kuhnian Relativism," Doppelt offers a reply to the above arguments. His reply, however, fails to make his original discussion more persuasive.

Doppelt allows that his version of incommensurability does not deny the logical or conceptual possibility that rival paradigms can engage in "coherent debate"; he claims only that in fact in particular cases there may be no room for such debate. Thus he denies that Kuhnian relativism holds that rational debate is in principle or *a priori* impossible. But this is a dramatic weakening of the Kuhnian incommensurability claim – which is exactly the point of my discussion (cf. this chapter, section B, 2). Doppelt thus weakens Kuhnian incommensurability from an "in principle" to a factual, contingent phenomenon. I have no quarrel with this weakened version of incommensurability, but I must reiterate that it is clearly inadequate to establish a conceptual relativism of the sort we are concerned with in this volume. The "in fact" incommensurability Doppelt points to does not yield a relativism of the sort depicted by ER; one can grant Doppelt's sort of relativism while accepting as compelling the self-refutation arguments against ER-type relativism. Doppelt writes that "the question at issue is not whether it is possible in principle for exponents of rival paradigms to share 'external standards' by which their debates might be governed." (120) But from the epistemological point of view this *is* the question. However interesting Doppeltian relativism might be, it is not the concern which has motivated the discussions of Protagoras, Plato, and subsequent philosophers concerning epistemological relativism. (Cf. further discussion of "moderate" relativism below.)

Doppelt denies that rational debate entails neutral standards and he holds that paradigm debate can be rational, and involve good reasons, in the absence of such standards. Doppelt's view is that such debate is rational in that "each side recognizes the empirical success of the other's problem-solutions as success, as a 'good reason' in the latter's favor." (120) But (a) this criterion – the empirical success of problem-solutions – is itself a neutral standard; moreover, (b) this fails utterly to render paradigm debate rational, for a paradigm need not recognize its rival's problems as *worth solving*, and so need not take successful problem-solution as a reason for embracing that other paradigm. Questions about the worth of rival problems and programs must themselves admit of rational adjudication if paradigm debate is to be properly construed as rational. Doppelt's use of scare quotes around "good reason" perhaps indicates that he is aware of the problem here: recognition of empirical success of problem-solution is not sufficient to render paradigm debate and evaluation rational. There must also be the possibility of neutral or non-paradigm-bound or non-question-begging discussion concerning the importance or worth of particular problems or solutions. But as Doppelt's discussion (120–121) indicates, on his conception of paradigm debate there is no assurance of such discussion, for proponents of rival paradigms can simply stay within their own paradigm, and not critically assess the merits of that paradigm with respect to its rival. Thus Doppelt has certainly *not* "account[ed] for the existence of rational debate between rival paradigms." (121) So long as paradigms are presupposed and protected from neutral evaluation, paradigm debate is as irrational as Kuhn's critics charge.

The general point is that paradigm debate must involve external, neutral standards by which rival claims and reasons can be judged if such debate is rightly to be taken as rational. If a proponent of a given paradigm is allowed to ignore the force of reasons against her view by retreating to internal standards which do not recognize those reasons as reasons, the debate cannot be regarded as rational. Consider Doppelt's Dalton example, according to which the pre-Daltonian has good reason to accept Daltonian theory, for he

can see that if he embraced the standards of the new Daltonian chemistry, certain achievements result which are lacking in his own research program. These achievements constitute good reasons for accepting the standards implicit in the [Daltonian theory]. Yet they need not be rationally compelling to the pre-Daltonian, because the different achievements made possible by his (pre-Daltonian) standards of chemical theory count as more important (better, more compelling reasons) than the Daltonian achievements relative to his standards. (121)

Here the Daltonian achievements are not genuinely recognized as good reasons by the pre-Daltonian, for the pre-Daltonian, on Doppelt's account, *can dismiss the force of these reasons*. As long as the pre-Daltonian (and, more generally, the proponent of a paradigm) can simply retreat to his internal, pre-Daltonian standards, and so avoid a neutral assessment of the merits of his paradigm with respect to its rival, it is incorrect to regard paradigm debate as rational, and to regard "achievements" such as Dalton's as *reasons* for embracing that paradigm. They must be regarded, rather, as "reasons," which the pre-Daltonian can ignore by retreating to his internal standards. Consequently Doppelt's analysis fails to secure the rationality of paradigm debate in the absence of neutral criteria by which rival paradigms and reasons for their acceptance can be non-question-beggingly assessed.[60]

In subsequent papers Doppelt develops and clarifies his defense of Kuhnian relativism. He is particularly concerned to distinguish between "extreme" relativism, according to which theory choice is irrational, and "moderate" relativism, according to which theory choice is rational in that it is based on reasons, but in which such choice is *under-determined* by reasons. His thesis is that Kuhnian relativism is most forcefully interpreted as a moderate relativism:

By "moderate relativism" I mean the claim that while there are typically some good reasons for theory change in science, there are often equally good reasons for adhering to the older theory. On this view, while scientific development is rational (there are typically some good reasons for the theory-changes which occur), new theories often fail to be demonstrably more rational than their predecessors, on any standards which were mutually acceptable and applicable to both. As a result, the thesis of moderate relativism is that scientific change is often or typically under-determined by good reasons.[61]

As this depiction of moderate relativism makes clear, Doppelt's Kuhnian relativism is not to be taken as an "in principle" or extreme relativism. Doppelt is not claiming that theory choice cannot, in principle, be based on good reasons because what counts as a "good reason" is relative. (This would be, in Doppelt's terminology, an extreme relativism.) His claim, rather, is that often, or typically, theory choice is based on reasons but is under-determined by those reasons: there is some reason to relinquish a currently held theory T_1 and adopt a new theory T_2, but there is also some reason not to make this change. Reasons in favor of neither theory are sufficient rationally to compel allegiance to it. (Accordingly, Doppelt seems to suggest that an account of science's rationality must have room for moderate relativism (absent a convincing case for rejecting that phenomenon); such an account must allow the possibility that any given episode of theory change is/was rational even though the change is/was under-determined by reasons.[62])

I argued above (section B, 2) that Doppelt's Kuhnian relativism, if taken as an *a priori*, "in principle" relativism, fails; in the immediately preceding paragraphs above I argued that if Doppelt interprets moderate relativism as a contingent, factual phenomenon – to wit, that in some cases of theory change reasons for the change under-determine the decision to change – then moderate relativism does not constitute a version of epistemological relativism as characterized in Chapter 1 by ER. As the preceding citation makes clear, it is in this second, contingent way that Doppelt understands moderate relativism.

Consequently, moderate relativism as Doppelt characterizes it does not constitute any sort of case for epistemological relativism or against the incoherence arguments against epistemological relativism. For moderate relativism is not a general thesis about the nature and possibility of knowledge, as epistemological relativism is generally taken to be. Perhaps an analogy with another general thesis about knowledge, skepticism, will be useful here.

Skepticism is generally regarded as a very general thesis about the nature and possibility of knowledge. The skeptic denies that we can know. Its philosophical bite comes with the claim that in *all* cases we *cannot* know. Skepticism wanes in philosophical interest if the skeptic claims merely that, in *some* cases we in fact *do not* know. To such a claim we can always reply, "let's find out; let's come to know". It is only when the skeptic suggests that we *cannot*, in principle, come to know that her thesis becomes philosophically problematic. Skepticism, to be forceful, must be taken as an "in principle" claim concerning the (im)possibility of knowledge.

Similarly, relativism loses its philosophical bite if the relativist claims (as does Doppelt's moderate relativist) simply that, in fact, in some cases of epistemic conflict there are not sufficiently powerful reasons to resolve the conflict. Its power comes with the claim that there *cannot* be such reasons. The moderate relativist makes no such claim; accordingly, her position is of little epistemological moment. To be forceful, relativism – like skepticism – must be taken as an "in principle" claim about the (im)possibility of rational resolution of certain sorts of disputes.[63]

So our preliminary conclusion (of section B, 5) stands: Doppelt has not advanced the cause of epistemological relativism. In the next chapter we will consider another Kuhn-inspired attempt to develop a cogent relativism in the context of a general Kuhnian, post-Positivist philosophy of science – that of Harold I. Brown.

CHAPTER 5

THE KUHN-INSPIRED NEW PHILOSOPHY OF SCIENCE

Harold I. Brown's articulation of the "new philosophy of science"[1] is particularly welcome because it systematically pulls together various strands of the post-positivist philosophy of science literature. The new philosophy of science Brown presents, drawing on the work of Kuhn, Toulmin, Hanson, Lakatos, Polanyi, and others, but effectively structured, enhanced, and defended by Brown, replaces logical empiricism's fundamental presuppositions of *Principia* logic and observation-based empiricism with a new approach which emphasizes the theory-ladenness of perception, the unavoidability of paradigmatic presuppositions, conceptual change and scientific revolutions, dialectic (as opposed to algorithmic) reasoning as the model of rational thinking in science, and a sharp rejection of the discovery/justification distinction. Brown develops a new epistemology of science to compliment and underlie the new philosophy of science, including alternative conceptions of rationality, scientific knowledge, truth, objectivity, and relativism. As with Kuhn, Brown's discussion of relativism cannot be helpfully separated from the rest of his presentation. Consequently, in what follows we shall concern ourselves with the entire Brownian conception of the new epistemology and philosophy of science. (In fact, while much of the discussion is indirectly related to relativism, this topic is explicitly addressed only briefly below.) Brown begins his discussion by analyzing the difficulties befalling the "old" philosophy of science – i.e., logical empiricism – and he develops his new philosophy of science in part by way of contrast with this older tradition. We begin, then, with Brown's analysis of logical empiricism and its relationship to the new philosophy of science.

1. THE LOGICAL EMPIRICIST RESEARCH PROGRAM – PRESUPPOSITIONS AND PROBLEMS

A central thesis of *Perception, Theory and Commitment* is that "there are fundamental similarities between scientific method and philosophic method."[2] In particular, Brown argues that both scientific research and philosophical research concerning science are endeavors controlled by presupposed theoretical frameworks, and that these frameworks of presuppositions guide the research efforts of both science and philosophy of science in the sense that they determine the problems that are to be solved, as well as the adequacy of proposed solutions to those problems. That all scientific research is guided by some presupposed framework is a central tenet of the new philosophy of science Brown defends. He also wants to argue that philosophical research is similarly framework-bound, and that the change from logical empiricist philosophy of science to the new philosophy of science "constitutes an intellectual revolution in philosophy of the same kind as scientific revolutions."[3] Brown seeks to expose, then, the presuppositional framework of logical empiricism, the way in which that framework determined the research program of logical empiricism, and the way in which failure to solve, by its own lights, its most pressing problems, led to the revolution in philosophy of science in which the framework of logical empiricism was overthrown, and the presuppositional framework of the new philosophy of science created in its place.

What are the presuppositions of logical empiricism? Tracing its development from Humean empiricism, Brown argues that two of logical empiricism's presuppositions were of fundamental importance in determining its research program: a commitment to the extensional logic of Russell and Whitehead's *Principia Mathematica*, and to some version of observation-based empiricism. Logical empiricism's commitment to these two presuppositions, according to Brown, determined both what problems were important research problems, and what could count as solutions to those problems.[4] The particular problems Brown focuses on to explicate his analysis are the problems of confirmation, theoretical terms, and explanation. In each case Brown tries to show that these problems are only problems given the logical empiricist's commitment to the presuppositions mentioned above. For example, Brown argues that the problems surrounding confirmation, especially Hempel's paradoxes of confirmation, stem largely from logical empiricism's commitment to extensional *Principia* logic with material implication. Similarly, the problem of explicating theoretical terms stems from logical empiricism's commitment to the thesis "that terms get their meaning as a result of having been correlated with some set of impressions or . . . 'sense-data'"[5] or observables, and flaws of various proposals to explicate theoretical terms in terms of observational terms (e.g., Carnap's attempted explication via reduction sentences) were deemed to be flawed by logical empiricism because of its commitment to *Principia* logic. Turning to explanation, Brown argues that difficulties with the D-N model of explanation again stem from the commitment to deductive extensional logic, and that this commitment determined both the sorts of solutions to the difficulties that could be countenanced, as well as those potential solutions which went beyond the pale of logical empiricism's presuppositional framework and so could not be accepted as solutions to the various difficulties recognized by the operative research program.

There is much of merit in Brown's account of the origins and development of logical empiricism. His discussion of the historical development of logical empiricism, tracing its roots through Hume, Wittgenstein, and Russell, is exceedingly clear and helpful. Nevertheless, there are difficulties with various portions of the analysis of logical empiricism's research program, and a general question concerning the epistemic status of presuppositions which will offer a convenient entry into Brown's presentation of the new, post-empiricist philosophy of science. Because our main concern is with that new philosophy of science, I will consider the difficulties with Brown's analysis of logical empiricism only briefly.

First, Brown chides the logical empiricists for consistently failing to analyze actual scientific theories and research, and instead for constructing artificial languages and calculi, illustrating these artificialities with simple empirical generalizations such as "All ravens are black." Brown complains that the assumption that such simple generalizations will illuminate the structure of science is unwarranted.[6] But in fact the empiricists were not seeking to elucidate the actual *structure* of science (whatever the "structure" of science might mean), but rather the *warranting relation* between evidence and generalizations. While the structure of actual science may be fantastically more complex than the simple generalizations that the logical empiricists were so fond of, the warranting relation between evidence and generalization, whether actual or simplified, is the same. In any case, it is this relation, and not the structure of science, that the logical empiricists hoped to illuminate with the help of their simplified examples and generalizations. The new philosophy of science Brown supports, we shall see, does not overly concern itself with this warranting relation, mainly because of its concern with the dynamics of scientific research and its emphasis on the scientific enterprise as a whole. However, we should not lose sight of the fact that the question of the epistemic status of scientific claims remains an important philosophical question, despite the new philosophy of science's tendency to lose sight of the question in its zeal to account for the dynamics of scientific change.[7] In any case, the present point is simply that it is not the structure of science, but the nature of the warranting relation between evidence and generalization, that the logical empiricists sought to get at with their simple examples.[8]

Second, Brown's account of Nelson Goodman's work on confirmation is in need of clarification. Brown argues that Goodman, is formulating his "new riddle of induction" and solving the riddle in terms of projection and entrenchment, in effect makes a major break with the logical empiricist research program, since determining the entrenchment of a predicate involves appeal to the facts of history. By resting the epistemic status of a given generalization on the facts of prior projection, Goodman gives up the logical empiricist project of providing a purely formal, syntactical account of confirmation. Indeed, Brown suggests that Goodman in effect is proposing a new research program altogether, by giving the history of science a central role in the determination of the confirmation of hypotheses by relevant evidence,[9] and that Goodman's approach to confirmation signals "that the standard form of logical empiricist philosophy of science has lost a good deal of its vitality as a research program and that a new approach may be in order.[10]

However, we again need to note that the primary focus of the logical empiricist approach is epistemological: it is the warranting relation between evidence and hypothesis that the literature on confirmation seeks to clarify. We must distinguish, therefore, between (1) the nature of the warrant evidence affords hypotheses, and (2) the establishment of any given hypothesis as *in fact* warranted by evidence. While the latter may involve excursions into the history of science (to establish the actual entrenchment of a predicate), the former need not. Establishing any given predicate as well entrenched is not part of the account of the warrant that (appropriately determined) evidence affords some given hypothesis (with entrenched predicate). To establish that any given hypothesis is in fact warranted, one of course has to appeal to factual information. This is as true for Hempel and Carnap as it is for Goodman. Brown is correct in pointing out that Goodman calls for a new sort of factual information to be relevant to the establishment of a given hypothesis as warranted, namely information concerning entrenchment. Nevertheless, establishing any given hypothesis as in fact warranted is not the same as accounting for the relation between a hypothesis and the evidence for it. Thus, while Goodman's work on induction and confirmation is clearly ground-breaking, it is not so clear that it represents as sharp a break with the standard logical empiricist approach to the problem of confirmation as Brown supposes.[11]

Third, in his discussion of explanation and truth, Brown unconvincingly discusses the nature of scientific truth as that notion figures in the logical empiricist literature. After attempting to show that contrary to the logical empiricist tradition, older laws (e.g., Galileo's law of falling bodies) are not strictly speaking entailed by newer laws (in this case, Newton's laws),[12] Brown argues that the logical empiricists were committed to the presupposition that "science does establish true propositions"[13] (in addition to the already mentioned presuppositions of *Principia* logic and an empiricism based on observables), and that in view of this commitment logical empiricists were bound to deny that older laws, such as Galileo's and Kepler's, have been shown to be false.[14] Brown argues that the logical empiricists were committed to the view that science produces results which "are true and stand forever," and that science provides "final truths."[15] This of course contradicts the empiricists' frequently made claim that empirical knowledge is probabilistic, fallible, and always subject to revision; Brown tries to show, therefore, that despite these disclaimers, the logical empiricists really did presuppose that science produces "final truths." In discussing the Hempel-Oppenheim treatment of explanation, Brown notes their distinction between explanations in which the explanans are true from those in which the explanans are only well-confirmed, explanations of the former sort being the only genuine sort of explanation, and argues that, in light of that distinction, "since they apparently hold that science does explain, they must also have held that science can get beyond a high degree of confirmation and discover premises which are *true, i.e., which can never be overturned*."[16] But here Brown simply confuses truth with certainty. To claim that *p* is true is not to claim that *p* is certain (can never be overturned). Rather, to claim that *p* is true is simply to claim that the world is as *p* says it is. True explanations – that is, explanations in which the explanans are true – are indeed achievable. This does not mean that those explanations are certain, or beyond revision. When we think we have good reason or a good case for some explanation, we take that reason as indicating truth. That is, truth constitutes the *import* of scientific claims.[17] There is nothing in the logical empiricist position which commits it to the view that science establishes certain truths. Indeed, Brown belies his confusion when he writes that one gets "beyond a high degree of confirmation" to discover "premises which are true,"[18] as if true statements are farther along on a continuum between falsehood and certainty than highly confirmed statements. True statements are not more certain than highly confirmed statements; rather, the import of highly confirmed statements is that they are true, or at least likely to be (approximately) true.

Moreover, Brown's discussion of logical empiricism's account of the cumulativity of true propositions is flawed because it fails to distinguish theoretical cumulativity from low-level empirical or experimental cumulativity. While logical empiricism may be committed to the cumulativity of data, it need not be committed to the cumulativity of theoretical propositions – thus the empiricist maxim that a new theory must "handle" all the data the old theory did in order to count as a legitimate successor. Even a new philosopher of science like Kuhn would agree here.[19]

In addition to these several objections to Brown's recounting of the specific machinations of logical empiricism's research program, there is a more general theme running through Brown's account of logical empiricism that demands attention. According to the new philosophy of science, all research programs are guided by presuppositions, and presuppositional frameworks determine the problems recognized by, and the adequacy of proposed solutions to those problems for, the research program in question. Brown extends this analysis to the philosophy of science, so that logical empiricism is also to be considered a research program, guided by its presuppositional framework of *Principia* extensional logic with material implication, observation-based empiricism, and the thesis that science does establish true (understood by Brown as certain) propositions. This framework, Brown has argued, helps explain the course of logical empiricism: the problems it focused on are identified as such by the framework, and the adequacy of proposed solutions to those problems is determined with reference to the framework. This raises an obvious question, namely, what is the epistemic status of the presuppositional framework? Are presuppositions criticizable? Capable of evaluation? Justifiable? For instance, is the logical empiricist justified in presupposing *Principia* logic? Can there be extra-framework bases from which to evaluate rival presuppositions, or are evaluations of presuppositions framework-bound? To answer these questions, we must see how presuppositions fit into the new philosophy and epistemology of science Brown constructs.

2. THE NEW PHILOSOPHY OF SCIENCE

The new philosophy of science has several strands. One is the unavoidability of presuppositional frameworks. Others are the theory-ladenness of perception; a view, not unlike Kuhn's, of scientific revolution; an account of the nature of discovery which rejects the discovery/justification distinction; and an emphasis on dialectical thinking in science.

Contrary to logical empiricism, Brown defends a strong version of the thesis that perception is theory-laden, and argues that the objects of significant perception are meanings.[20] Theories and concepts guide perception. But theories play a much more basic role in science: they guide the research efforts that constitute the scientific enterprise. Fundamental theories are presupposed by research scientists; they form the presuppositional framework of normal science, "in which an accepted fundamental theory serves to organize and structure scientific research. The theory determines the meaning of observed events by providing the scientist with grounds for recognizing which observations are relevant to his research, which of the relevant ones pose problems, how the problems are to be attacked, and what counts as an adequate solution to a problem."[21] Brown puts the point even more forcefully as follows:

> In order to carry out meaningful research we require a research problem and some criteria of what evidence is relevant to its solution. More fundamentally, we require some basis for deciding what research problems are worth pursuing. It is our accepted theories, the systems of presuppositions to which we are already committed, which provide this basis. And because we always do research within a system of presuppositions, both our problems and our data are thoroughly theory-laden.[22]

Propositions which are presuppositions form, moreover, an "epistemically distinct class" of propositions which Brown labels, after Kuhn, "paradigmatic."[23] Paradigmatic propositions are not analytic or a priori. But neither are they "ordinary empirical propositions," because they are "protected from straightforward empirical refutation."[24] We might note that this property of paradigmatic propositions does not itself serve to distinguish such propositions from ordinary empirical propositions, since, given Brown's commitment to the theory-ladenness of perception, even ordinary empirical propositions are protected from straightforward empirical refutation. Once theory-ladenness is taken seriously, straightforward empirical refutation is impossible. But this is a minor criticism. The point Brown is concerned to make is that paradigmatic propositions are those "which come to have a constitutive role in the structure of scientific thought."[25] (Examples of paradigmatic propositions are: "celestial motions are circular"; "physical space is Euclidean.")

Despite the constitutive role of paradigmatic propositions in the structure of scientific thought, and despite also their immunity from straightforward empirical refutation, these propositions, and the presuppositions they express, are capable of being abandoned or revised. Following Kuhn, Brown argues that the research program defined by a presuppositional framework occasionally turns up anomalies, and that an anomaly sometimes turns out to be a "genuine counter-instance"[26] which forces the research community to alter (at least) part of its presuppositional framework. The fact that "the search for truth"[27] can make researchers realize that one of their presuppositions is false raises a difficulty Brown. For he holds both that

(1) Presuppositions organize and structure scientific research, and

(2) Research may force the abandonment (as false) of presuppositions. Presuppositions are fallible.

The difficulty is: are these two claims compatible? It seems, *prima facie*, that they are not. For, as we have seen, presuppositions structure scientific research in the important sense that they provide criteria of adequacy by which we evaluate proposed solutions recognized as such by the research program delineated by a presuppositional framework. Thus, presuppositions determine how we evaluate evidence for or against hypotheses being investigated. But if presuppositions determine not only what counts as evidence, but also the evaluation of the evidence, it is difficult to see how evidence could cast negative light on a presupposition – for if it can, in what sense can the presupposition be said to determine the evaluation of evidence? If (1) and (2) are not outright incompatible, (2) must at least greatly limit the force of (1), since the scientist can call into question and evaluate (according to presupposition-neutral criteria of evaluation of presuppositions?) the presuppositions that organize and structure, and so determine, the scientists' research program. Indeed, it is not clear what, on Brown's view, a *genuine* counter-instance to a presupposition can be, since what counts as an instance is determined by the very presupposition countered.

The general question raised here, then, is this: given the overarching role of presuppositions in structuring scientific research, how is it possible for such research to offer evidence which forces the abandonment (as false) of these very presuppositions? Does that possibility not necessitate the existence of framework-neutral criteria of evaluation of presuppositions by which frameworks can be evaluated? It is not clear how it can be held that evidence can force the abandonment of presuppositions without weakening the power of the presuppositions to determine the evaluation of that evidence.[28] In any case, normal science is research guided by a presuppositional framework, and scientific revolution, the next aspect of the new image of science to be considered, is understood by Brown as the transformation of that framework.

Brown's account of scientific revolutions, like that of normal science, closely parallels Kuhn's. For Kuhn, a scientific revolution is the overthrow of a paradigm by a new one. For Brown, a scientific revolution is the transformation of the presuppositional framework (expressed by paradigmatic propositions) guiding normal science into a new framework. In addition, the world, understood as the "meaning structure" of the scientist,[29] is altered during a scientific revolution. Concepts are transformed during scientific revolution, so that revolutions are the loci of conceptual change. In his admirably clear extended discussion of the Copernican revolution, Brown identifies several presuppositions that were successfully challenged, especially the distinction between terrestrial and celestial physics and the nature of "natural" motion in the two spheres, and illustrates the way in which the transformation of presuppositions transforms research programs by solving or eliminating old problems and creating new ones. His central concern, though, in his discussion of scientific revolutions, is not the transformation of presuppositions but of concepts.

Conceptual change occurs typically during scientific revolutions, according to Brown. Both the sense and the reference of concepts are altered. Brown illustrates this claim in his discussion of the concept "planet." For the pre-Copernicans, the sense of "planet" included the requirements that it move around the earth and in relation to the fixed stars, while after Copernicus motion around the sun became a defining characteristic.[30] Moreover, the reference of "planet" also changed, since the earth was included in the extension of the term after Copernicus, but not before. Likewise, the concepts "sun" and "star" have changed both in meaning and in reference: whereas now the two are synonymous, for pre-Copernicans the sun was unique. Brown argues that "after a revolution we find recognizable modifications of old concepts,"[31] so that, for example, while both the sense and reference of "planet" were changed during the Copernican revolution, "many aspects of both remained the same throughout this change. For example, planets are still celestial objects which move with respect to the fixed stars, and there is considerable overlap between the extension of the old concept and of the new one."[32] In fact, Brown sees successive changes in a concept across revolutions as continuous and gradual. Borrowing a metaphor from Hempel, Brown offers the following pictorial account of concepts and conceptual change:

> A scientific concept is a knot in a web; the strands in the web are the propositions that make up a theory; the meaning of a concept is its location in the web. Thus the meaning of a concept is determined by the strands that come into this knot, by the other knots the one in question is connected to, and by the further connections of these other knots . . . . a concept is not a simple which is grasped entirely or not at all, but rather a complex which can be learned a bit at a time.
>
> . . . we can apply the web metaphor to concepts that are transformed as a result of a scientific revolution. Some of the strands which come into a particular knot are removed, others are redirected, and some new strands are introduced. The concept retains some of its old characteristics since some of the old strands are left intact, but it also loses some old relations and acquires new ones, thus we acquire a new version of an old concept.[33]

There is much that recommends this picture of conceptual change. First, it is a gradualist account in that it holds that concepts change slowly over time in complex ways, and change in ways that allow new concepts to retain much of the meaning of their conceptual ancestors. This gradualism and continuity avoids problems which plague other accounts of conceptual change offered by holders of the new philosophy of science, many of which revolve around some version of the incommensurability thesis. In addition, Brown's account of conceptual change does considerable justice to the variety of instances of conceptual change he considers. Nevertheless, there are some difficulties with the account. First, it is unrelievedly metaphorical. Second, it is not clear that Brown's account of conceptual change is an improvement over the more traditional account of theory change according to which it is a *recategorization* of the world, together with a reassessment of propositions describing the world, rather than reconceptualization, that occurs during revolutions. Consider Brown's example of the concept "planet." Brown argues that that concept's sense and reference changed during the Copernican revolution: its sense was altered, so that while before Copernicus its sense included the requirement that a planet revolve around the earth, after Copernicus its sense included the requirement that it revolve around the sun. Now, Brown holds that this is a case of the sense of the concept "planet" changing. But we might as plausibly construe this as a case in which a hypothesis concerning a feature or property of planets – that they revolve around the earth – is reassessed and given up; and an alternative hypothesis – that planets in fact revolve around the sun – adopted in its stead. In other words, it is not clear that we have here a change of concept, or rather a change of estimation of a proposition or hypothesis couched in terms of some concept. But any adequate account of conceptual change must be able to draw this distinction. Consider next Brown's claim that the reference of "planet," as well as its sense, changed during the Copernican revolution, so that while before the revolution the earth was not in the extension of the concept, after the revolution it was. We may, as Brown does, consider this a change in the reference of "planet." But we may also regard this not as a *conceptual* change, but as a recategorization in which the object Earth is now sorted into the category planet, whereas before it was not – the basis of the recategorization being a revision in our estimation of hypotheses concerning the properties of planets. Brown is surely correct that for early astronomers, Venus, Mars, and Jupiter, but not Earth, were included in the extension of "planet"; while for post-Copernicans Earth was included in that extension as well. But, if what occurred during the Copernican revolution was (in part) a re-classification of heavenly objects, based on a new view of the properties of planets in which planets were held to revolve around the sun instead of the earth, then we might well regard this as a change, not of the concept planet, but of the principles of categorization according to which the earth is now seen as belonging to that category. Before the revolution, the world included the same objects – the usual planets, plus the earth – as after the revolution. But because the new classification scheme sorted these objects differently, the earth was recategorized as a planet.[34] This view of the change in which the earth was taken to be a planet recognizes the distinction between change of concept, and change of estimation of a hypothesis couched in terms of some concept. Moreover, recognizing that distinction allows us to raise the most important question, epistemologically speaking, concerning conceptual change: how are such changes evaluated and justified? The third and most crucial difficulty with Brown's account is that, according to it, there seems to be no way to justify conceptual change.

On the traditional view of theory change sketched above, it is clear how one goes about justifying such change. New categorizations may be justified, for example, in terms of increased explanatory power; or in terms of evidence for hypotheses formulatable in the idiom of the new categorization; or on pragmatic grounds; and so on. However, these sorts of justificatory moves regarding conceptual change are not available on Brown's account, for cases of conceptual change on that account are changes in the meanings (senses and references) of concepts, not changes in propositions couched in terms of those concepts. Justificatory tactics such as explanatory power, increased evidential support, and the like, apply to the latter, not the former, sort of change. To take Brown's example: how do we (Copernicans) justify the shift in the concept planet, in which planets are taken to revolve around the sun, and the earth is counted as in the extension of the term? Presumably, we do so by noting that there is evidence that the earth is in fact a planet in that it has central properties that other planets have; that a system of propositions which is couched in terms of the new conception of "planet" (i.e., a portion of astro-physical theory) has more explanatory power than the system of propositions couched in terms of the older conception of "planet"; that there are pragmatic advantages (of prediction, for example) to using the new conception; and so on. It is crucial to see that these justificatory tactics apply to propositions couched in terms of concepts, not to concepts *simpliciter*. For when we seek to justify – that is, when we seek good reason for accepting or rejecting some claim or concept – we have no choice but to look to propositions, since concepts, taken to be simply "knots" of senses and references, isolable from propositions and claims which utilize, or are couched in terms of those concepts, are incapable of justification. To ask the epistemological question, "which conception of 'planet' should we accept?," is to ask which set of propositions, utilizing one or the other conception, can be justified.

The point, at bottom, is this: it is claims for which justification can be sought; not concepts. To understand the epistemic features of conceptual change, therefore, we must construe conceptual change not simply as changes in meanings, but as changes, as it were, in the theoretical edifices constructed with the building blocks of alternative concepts – that is, as changes in claims couched in terms of alternative concepts. Thus Brown's account of conceptual change, by focussing too narrowly on the meanings of concepts, fails to face the important epistemological problems revolving around that phenomenon.[35] This is ironic, since Brown argues forcefully for the interconnectedness of concepts and propositions in scientific theorizing.[36] It is this very interconnectedness, nevertheless, that undermines his "knot" theory of conceptual change, for that theory fails to help us understand the epistemic features of conceptual change precisely because it construes change exclusively in terms of changes in the meanings of concepts.[37]

Placing the emphasis on scientific revolutions that he does in his presentation of the new image of science, Brown naturally raises questions concerning the development of new hypotheses, concepts, and theories, which in turn raise questions concerning the nature of scientific discovery. Given the legacy of positivism, one can hardly consider scientific discovery without considering Reichenbach's distinction between the context of discovery and the context of justification. As did Kuhn, Brown argues, not unpredictably, against that distinction. But here his discussion is flawed.

Brown argues that "there are two theses built into" the discovery/justification distinction: "that a sharp line can be drawn between the discovery and the testing of scientific theories; and that it is only with respect to the testing of theories that we can speak of a logic at all, for logic has nothing to say about discovery."[38] Brown argues that these are dubious; we will look at those arguments in a moment. At the outset, however, we must note that these two theses, even if they are "built into" the discovery/justification distinction, fail to capture the central thrust of the distinction, which is at bottom epistemic: namely, that facts concerning the discovery of hypotheses or theories cannot contribute to the justification or epistemic warrant of those hypotheses. The heart of Reichenbach's distinction is that discovery is *irrelevant* to the evaluation of the epistemic worth of (scientific) knowledge-claims.[39] Since Brown's arguments against the distinction fail to come to grips with this aspect of it, those arguments are in an important sense beside the point. Nevertheless, it is instructive to look at Brown's arguments against the two theses, since his account of scientific discovery and change hinges in part on those arguments.

The first thesis Brown wants to reject is that a sharp line can be drawn between the discovery and the testing of scientific theories. Here Brown argues that such a line cannot be drawn, because we do not usually refer to a proposal as a discovery unless it has passed enough tests to become . . . a part of the accepted body of science . . . . When we credit Galileo or Newton or Einstein or Bohr with having made scientific discoveries, we only consider those hypotheses which they had good reasons for entertaining to be discoveries. The context of justification is thus part of the context of discovery and no sharp line can be drawn between discovery and justification.[40]

There are a number of points to be made against this argument. First, it is, contrary to Brown, the invention of a new hypothesis or theory, prior to testing, that adherents of the distinction refer to as "discovery." Thus it is a straightforward equivocation for Brown to count as discoveries proposals which have "passed enough tests to become . . . a part of the accepted body of science." We must distinguish between two senses of discovery: that which refers to the conceiving of a new hypothesis, and that which refers to those hypotheses which have already been put to, and have passed, the test. It is clear from the passages Brown cites[41] from representative logical empiricists that it is the former sense of "discovery" they are concerned with, and that they are concerned to argue that discovery, so understood, cannot contribute to the epistemic evaluation of the hypotheses discovered. Thus Brown's argument against the first thesis equivocates, and fails to come to grips with the main point that the logical empiricists wish to make with the use of the distinction. Moreover, Brown's argument confuses, and makes it impossible to distinguish, various "cognitive stances" we can take toward hypotheses and theories. We want, for instance, to be able to distinguish between theories we reject, theories we are agnostic about, theories we think are worthy of further pursuit, theories we accept as plausible, theories we take to be true, and so on. The careful delineation of such cognitive stances is of the utmost importance, both epistemologically and methodologically; but Brown's view of discovery obscures them, since the only discoveries he recognizes are those which become "a part of the accepted body of science."[42] Brown's argument against the first thesis thus neither effectively rebuts Reichenbach's distinction, nor establishes his own account of discovery as either plausible or helpful.

Brown's argument against the second thesis – that logic only applies to testing hypotheses, not discovering them, so that discovery must be arational – is more successful, but again fails to come to grips with the logical empiricist's distinction between justification and discovery. Brown argues here that there is a role for reasons in the organized attempt to discover plausible hypotheses and theories. Here Brown is surely right: as Peirce's "abduction" and Hanson's work on the logic of discovery attests, there are clearly maxims or rules of thumb which investigators may use in the effort to discover.[43] Nevertheless, as I have argued elsewhere,[44] the fact that there is room for reasons in the organized effort for discovery does not in and of itself challenge Reichenbach's distinction, since such reasons (or maxims of discovery) must be independently shown to have justificatory force. One need *not* hold that the process of discovery is arational to hold that discovery offers no justification. In any case, having argued against the thesis that logic has no role to play in discovery, and decried the positivist identification of rationality with logic,[45] Brown offers an account of reasoning during the process of discovery which exhibits the way in which discovery can be rational. This account of reasoning Brown labels "dialectical."

Dialectical reasoning takes place "in the context of a set of presuppositions which provide criteria for determining what constitutes an acceptable answer"[46] to a given problem or question. Questions "only arise in the context of presuppositions,"[47] and presuppositions provide criteria for the evaluation of answers. However, dialectical reasoning can also call into question various presuppositions, and so challenge operative criteria of evaluation. Dialectical logic is not a formal, but rather a "content logic" which constitutes for Brown "a tool for examining the structure of research in terms of the historical context."[48] This focus on context, Brown argues, is necessary, for "it is impossible to understand the structure of scientific research without understanding its content in its historical setting."[49] Brown offers several examples[50] to illustrate dialectic in scientific discovery, showing how even the most bold scientific revolutions began with the scientist working within the problem situation and presuppositional framework of the soon to be overthrown theory. Thus Copernicus accepted the problems left him from the Ptolemaic tradition, and Einstein those of classical electrodynamics. In each case, Brown argues, the revolutionary developments are dialectical:

"Beginning from an existing theory, i.e., an answer to a question, Einstein gives reasons why this answer is inadequate and proposes a new solution, one which may be quite different from the one he rejects but which nonetheless takes its departure from some of the ideas of the rejected hypothesis. Just as there is no logically necessary reason why Einstein must abandon classical theory at this point rather than joining the efforts to make it work, so there is no logically necessary reason why he must seek a new approach in the particular way that he does. Still, Einstein consciously built his new theory on a foundation provided by the classical physics he was overturning."[51]

We could substitute "Copernicus," or the name of any other scientist whose work facilitated a scientific revolution, for "Einstein" in this passage. The point Brown wants to emphasize is the dialectical relation between old and new theory; the way in which the new "grows out of" the old, while taking as its starting point the problem situation and presuppositional framework of the older tradition. In this way, scientific revolutions can be seen to be eminently rational: not logically necessary, but dialectically compelling, in which problem situations and presuppositions offer continuity and evaluative standards even as they are piecemeal changed or reconceptualized.[52]

The continuities between successive theories offer standards for evaluating instances of theory change. These standards are not "eternal" standards erected by philosophers but rather scientific standards "provided by the theories involved."[53] For example, Maxwell's equations provided a scientific standard by which to compare pre- and post-Einsteinian physics; the equations constitute "a standard internal to physics which serves as a point of mediation in this dispute."[54] Discussing Galileo's dispute with the Aristotelians over the importance of the standard of quantitative accuracy, Brown argues that here too "the contemporary state of physics provides criteria for deciding between two competing fundamental theories although these are not universal criteria accepted independently of the history of science."[55] Generally, Brown argues that "in a dispute between two radically different physical theories, there are a number of specific phenomena that the proponents of both theories accept as relevant and which can serve as the basis for rational debate. But which phenomena are accepted and what criteria are applied in evaluating the theories are determined in the specific problem situation, not by universally valid a priori considerations."[56]

There is one clear advantage of Brown's account: emphasizing as it does the continuity between even radically divergent theories, it avoids the difficulties raised by non-continuous accounts of theory change, especially incommensurabilist accounts such as Kuhn's. Nevertheless, Brown's discussion of standards and dialectic faces severe difficulties. Let us first look at his account of standards. As noted above, for Brown standards are not eternal and a-historical, but rather are provided by the theories involved in cases of theory dispute. Thus, for example, Maxwell's equations, though having different meanings in the context of the two theories, constituted a standard for comparing pre- and post-Einsteinian physics, since both physical theories accepted the equations.

If both theories indeed accept the equations, a problem arises – namely, how can the equations serve as a basis for *evaluation* of the theories? Standards of evaluation must *differentiate* between rival theories, or at least it must be possible for them to so differentiate. But on Brown's account, internal standards such as Maxwell's equations cannot possibly serve to differentiate between the two theories, since they count as standards precisely because both theories accept them. If we ask, "which theory better satisfies the standard of Maxwell's equations?", we cannot help but answer, "both equally satisfy that standard, since both accept those equations." We cannot argue, for example, that Einsteinian physics better meets the standard in that, for example, it better explains the equations, for that would be to bring in non-internal standards (of adequate explanation) that Brown is most concerned to deny with his claim that standards are internal and scientific, and not the eternal, a priori standards invented by philosophers. Brown's standards thus are incapable of functioning as evaluative standards. They are, rather, marks of continuity; they are indeed capable, as Brown argues, of providing "touchstones for debate" in that both sides in the dispute (by hypothesis) accept them. But Brown confuses items shared by disputants with items that can serve to evaluate differentially the views of disputants. The very fact that Brown's internal standards are shared by the disputants renders them impotent to serve as standards of evaluation, and so those internal standards cannot be taken, as Brown takes them, as evaluative standards capable of establishing the epistemic merit of rival theories.

This leads us to a more general difficulty with Brown's account of standards. He argues, recall, that, for any given theoretical dispute, "what criteria are applied in evaluating the theories are determined in the specific problem situation, not by universally valid a priori considerations.[57] We must distinguish between two questions about criteria of evaluation we can ask, a distinction blurred in the passage just cited: for a given dispute, we must distinguish between (1) what criteria are (were) *in fact* applied by the scientists in question, and (2) what criteria should be (have been) applied – that is, what criteria offer *epistemic warrant* for accepting one or the other theory. Suppose, for example, that Einstein's theory was accepted by the physics community because it met standards such as Maxwell's equations. (As argued above, this is impossible, since those equations were accepted by both theories – otherwise they would not, on Brown's account, constitute a standard – and so cannot evaluatively distinguish the two theories. But let us ignore this difficulty for the moment.) We may well want to ask further questions about that acceptance. For instance, we may want to ask why these standards should be taken seriously; why meeting these standards provides good reason for accepting Einstein's theory. That is, we want an account of the epistemic warrant of the theory; an account which tells us what the epistemological significance is of the theory's meeting some standards. Philosophers typically do this by talking of connections between external standards of evaluation – for example, explanatory power, simplicity, fertility, etc. – and epistemic properties of theories, such as warrant. By limiting his standards to internal ones (which, as argued earlier, is impossible if standards are to evaluate), Brown eliminates the possibility of even raising questions of epistemic warrant – standards are standards not because of their epistemological importance, but because the disputants in a theory conflict accept them. This is surely inadequate as an account of evaluative standards, unless we are willing simply to stop worrying about the epistemological underpinnings and significance of science.[58]

Finally, Brown ignores, in his rejection of external standards, the standard standards philosophers take to be most important: for example, those of explanatory adequacy, scope, fertility, simplicity, and so on. It will not do simply to reject these as standards because they are (if they are) "eternal," a-historical, or a priori. Since these are the standards philosophers of science typically consider to be important, there is an important sense in which Brown's discussion fails to join the issues with the older philosophy of science he seeks to replace.

Brown's discussion of dialectic, as we have seen, has the great merit of emphasizing continuity across scientific revolutions, in that even the most radical developments dialectically grow out of earlier presuppositional frameworks and problem situations. Nevertheless, the account of dialectic Brown offers does not bear the weight he places on it. It is too vague and simple to bear that weight. It does not tell us, for example, when a potentially revolutionary development dialectically evolves, how we tell whether the proposed development is a justifiable one or not. Or again, when in dialectic a presupposition is challenged, Brown's account cannot illuminate us as to when it is appropriate to drop the presupposition, and when not. Brown is surely right that scientific revolutions are not logically necessary. But we do not know yet how to tell whether dialectical developments are rationally compelling or not, for we have no criteria by which to evaluate those developments – except the internal scientific standards Brown offers, which, as I have argued, cannot do the job. Without such criteria, dialectic allows too much to be counted as rationally compelling, and offers no way to differentiate between those dialectical developments which are justifiable and those that are not; those "growths" that are legitimate and those that are not. These difficulties with dialectic are further clarified by their relation to Brown's conception of rationality. That conception is developed in Brown's presentation of the new epistemology of science, which is "implicit in the new philosophy of science."[59]

3. THE NEW EPISTEMOLOGY OF SCIENCE

Brown rejects the conception of rationality offered by the logical empiricists – for whom, according to Brown, rationality is identical with algorithmic computability[60] – on grounds that, at least in science, rational decisions concerning, for example, "how a discrepancy between theory and observation is to be handled requires a judgment by scientists. The decision cannot be made for them by the simple application of an algorithm."[61] Generally, Brown argues, "informed judgment" is required of rational decisions and deliberations; cases "in which we must rely on human judgment" constitute paradigm situations in which rationality or reason is required.[62] Informed human judgment thus constitutes a necessary condition of rationality. Brown here draws a parallel between his account of rationality and Aristotle's account of deliberation in ethics, in which Aristotle's "man of practical wisdom" is analogous to Brown's rational scientific decision-maker. Neither rely on algorithms; the conclusions of the judgments of both are fallible; for both, qualified judgment requires experience and knowledge. Brown argues that in Aristotle's man of practical wisdom we have "a model of the maker of crucial scientific decisions which cannot be made by appeal to an algorithm, and I offer the making of these decisions as a model of rational thought. It is the trained scientist who must make these decisions, and it is the scientists, not the rules they wield, that provide the locus of scientific rationality."[63]

There is much in Brown's account of rationality which is welcome. In particular, his emphasis on the requirement of informed human judgment in rational decision-making is a clear improvement over the identification of rationality with algorithmic computability. Nevertheless, the account of rationality offered is not, as it stands, satisfactory. For while informed human judgment may well be a necessary condition of rationality, it is surely not a sufficient condition. Brown's account counts too much as rational, and offers no way to distinguish rational judgments from irrational ones.

Again drawing on Aristotle, Brown clarifies his conception of rationality: for Aristotle, the man of practical wisdom recognizes that justice sometimes requires the violation of universal law; similarly, for Brown, rational scientific judgment requires the violation of, for example, methodological rules. So, for example, while scientists may accept the methodological rule to reject a theory if it is inconsistent with a well confirmed falsifying hypothesis, in certain episodes of the history of science – Leverrier's declining to reject Newtonian theory despite Uranus' failure to conform to the theory's predictions;[64] or Galileo's declining to reject Copernicanism despite contradictory evidence[65] – rational judgment requires that methodological rules be violated or suspended. This is fine as far as it goes; Brown is on firm ground in holding that it was rational for scientists, like Leverrier and Galileo, to ignore methodological rules in certain instances. But surely in some cases it is irrational to ignore such rules. How do we differentiate between these two cases? How do we know, for instance, that while Galileo's decision to continue to hold Copernican views in spite of the evidence was rational, the flat-earther's decision to hold the view that the earth is flat in spite of the evidence is not? We need to know when it is *justifiable* to hold on to hypotheses in the face of contradictory evidence, and when not. Brown's account offers no criteria by which we can assess the justifiability of decisions. All informed human judgments, on his account, are rational. But this is false – many informed human judgments are irrational – and a satisfactory account of rationality must allow us to differentiate between those judgments which are rational and those which are not. So while Brown is correct in viewing rationality as a function of judgment, not algorithmic computability, this insight does not by itself constitute a satisfactory account of rationality.

Rational judgments, for Brown, are fallible. This fact – that judgments can be rational, yet false – leads Brown to offer a new analysis of scientific knowledge, for, he argues, the old epistemology of science, held by, among others, the logical empiricists, construed scientific knowledge as infallible.[66] To "free" the concept of scientific knowledge from the concept of infallibility, Brown simply construes scientific knowledge as that which the scientific community takes to be knowledge: "scientific knowledge in any era is what the scientists actively take as such, and the scientific knowledge of one era may be rejected as error in the next. But the rejection of previously accepted claims will itself be made on the basis of the currently accepted views, which are themselves fallible."[67]

It is surely a plus for Brown's analysis that, on it, scientific knowledge is fallible.[68] Nevertheless, there are severe drawbacks to his account. Most obviously, on Brown's account there can be no distinction drawn between scientific knowledge and belief, since knowledge is simply what scientists believe in any given era. This is, in effect, not to provide a new account of scientific knowledge, but to give up on the task of providing one. Without a way to distinguish beliefs which count as scientific knowledge from those which do not, we have no account of scientific knowledge. At best we have an account of scientific belief: to wit, "scientific belief in any era is what the scientists of that era believe." Brown might try to draw the distinction between beliefs which count as knowledge and those which do not by appeal to the scientists' rationality, but such an appeal cannot succeed, since, as we saw earlier, Brown's account of rationality fails to distinguish those judgments which are rational from those which are not. Since that account also fails to provide any criteria by which we may evaluate the justifiability of judgments, any appeal by Brown to a distinction between justifiable and unjustifiable beliefs in an effort to distinguish beliefs which count as scientific knowledge from those which do not is likewise doomed to failure.

As Brown recognizes, his account of scientific knowledge is relativistic. Brown accepts the charge, arguing that this is the only way we can avoid an absolutist, i.e., infallibilist, epistemology of science.[69] Here Brown's confusion of truth and certainty, noted earlier, comes home to roost. According to Brown, "anti-relativism amounts to a claim that only propositions which are true, *in the ultimate sense*, [i.e., *certain*] can be a part of science . . . . Unless scientists have an effective method for determining *once and for all* which propositions are true, we cannot determine which part of currently accepted science is indeed knowledge."[70] This view of anti-relativism, as well as the view that truth "ultimately" is equivalent to certainty, is straightforwardly false. Anti-relativism does not demand certainty, or ultimate truth, but only that knowledge-claims be evaluated non-arbitrarily, and that propositions be more or less warranted on the basis of reasons which serve to justify their acceptance.[71] Moreover, to construe scientific knowledge's truth as certainty is, as I argued earlier, to mistake the referential import of truth. To claim that a scientific proposition is true is not to claim that is it certain; rather, it is to claim that the world is as the proposition says it is. It is simply false that a non-relativistic account of scientific knowledge must construe that knowledge as certain. All that is required is that propositions be judged non-arbitrarily, and that propositions that are accepted have some sort of warrant which justifies that acceptance; the truth of such items of scientific knowledge lies not in their certainty but in their referential import – that is, in their appropriateness as assertions about the world.[72]

Brown argues that in his account of scientific knowledge such knowledge is not arbitrary, since it depends on the rational (informed) judgment is scientists. This does not save his account from arbitrariness, however, since his account of rationality, as we have seen, offers no way to distinguish, from the set of informed, qualified judgments, those which are rational from those which are not. Brown argues further that his acceptance of relativism is innocuous, since it only amounts to fallibilism. Here, however, he is mistaken; because he fails to provide any way to assess the warrant of knowledge-claims, or the justifiability of their acceptance, his position is not just relativistic in the weak sense in which scientific knowledge is fallible. It is relativistic in the much stronger sense that scientific knowledge-claims are arbitrary, in that there is no way to evaluate such claims. Brown's appeal to the rationality of scientists is no help, since on his account of rationality there is no way to distinguish warranted from unwarranted beliefs. Thus Brown's criticism of infallibilist epistemology[73] is simply an attack on a straw man. For the old epistemology need not be committed to scientific knowledge's being certain, but only true; and in any case there is no reason to think that our only options are to embrace either infallibilism or relativism. Brown argues that we should accept the latter because we must reject the former, but we can reject infallibilism without accepting relativism, by providing an account of scientific knowledge in which that knowledge is not certain or infallible, but only warranted, and in which truth is not taken "ultimately," as certain, but rather as an indication of the referential import of scientific knowledge-claims.

Brown's acceptance of relativism leads to further difficulties with his conception of truth. He argues, rightly, that truth cannot serve to evaluate theories – that is, we do not choose T_1 over T_2 because T_1 is true. He concludes, wrongly, that since truth cannot serve as a tool of evaluation of theories, it is impossible to "discuss meaningfully the truth or falsity of actual scientific theories."[74] (This is wrong because we can still discuss meaningfully the truth or falsity of actual scientific theories, when we consider their epistemic import. Evaluative tools are not the only "meaningfully discussables" of science.) Giving up on truth as correct description of reality (which Brown labels "$truth_1$"), Brown introduces another sense of truth, "$truth_2$", which is compatible with his relativistic view of knowledge: "any proposition which is part of the body of scientific knowledge is a $true_2$ proposition."[75] There are difficulties with this account of truth. First, a claim is $true_2$ if it is accepted by the scientific community. Thus to say of some proposition that it is $true_2$ is simply to say that it is believed by the members of that community. The distinction between true and false belief is lost. Perhaps more importantly, $truth_1$ on Brown's account becomes irrelevant to the epistemology of science. Brown acknowledges that correct description of reality is "the goal for which scientists strive in constructing theories."[76] But, given his account of truth, in which only $truth_2$, and not $truth_1$, plays a role in the epistemology of science, the overarching goal of theorizing plays no epistemic role. Given that goal of theorizing, however, we may wonder why $truth_2$ should play any role in the epistemology of science. What is the point of $truth_2$ if the epistemic goal is $truth_1$? Brown does not claim any connection between $truth_1$ and $truth_2$ – for example, he does not claim that $truth_2$ counts as evidence for $truth_1$. Thus we are in the uncomfortable situation in which what we are after is $truth_1$, but our epistemology can only deal with $truth_2$. All the deep questions concerning the $truth_1$ of scientific claims – what is it for a claim to correctly describe reality?; what counts as evidence for the truth of a claim?; (how) can we tell if scientific theorizing is approaching its goal of correctly describing reality? – are simply not addressed by Brown's analysis. His account of truth then does not provide an analysis of scientific truth; it rather gives up on the very possibility of such an analysis. And in any case, as we have seen, it fails to come to grips with the referential import of (true) scientific propositions.

We must, finally, consider Brown's account of the objectivity of science, particularly as it ties in with his discussion of presuppositions. The reader will recall that, according to Brown, all scientific judgments require presuppositions. Since all judgments require presuppositions, and all presuppositions are fallible, objectivity cannot be a matter of resting on infallible foundations. In what, then, does scientific objectivity consist?

Brown is surely right in claiming that objectivity does not consist in tying scientific claims to infallible foundations. He adopts a view of objectivity that is, in fact, very close to the standard "good reasons" view – namely, that objectivity consists in the impartial evaluation of considerations weighing for or against a theory. So, for example, objective grounds for accepting a theory are that "it solves some problems, eliminates others, and provides a guide to further research."[77] This account of objectivity is, in my view, while only briefly developed, on the right track. The problem is that it is incompatible with Brown's account of presuppositions.

Recall that, for Brown, a presuppositional framework "determines the meaning of observed events by providing the scientist with grounds for recognizing which observations are relevant to his research, which of the relevant ones pose problems, how the problems are to be attacked, and what counts as an adequate solution to a problem."[78] Thus there are, according to Brown, no framework-neutral criteria for evaluating the adequacy of proposed solutions to problems. Criteria of evaluation are framework-bound. But in his account of objectivity, Brown appeals to framework-neutral criteria: problem-solving ability; problem-eliminating ability; and research guidance ability. These clearly function as framework-neutral criteria since, Brown argues, a theory may be objectively rejected when "an alternative theory has been developed which is judged to meet these [just cited] criteria."[79] Brown's account of objectivity thus depends on the existence of presuppositional framework-neutral criteria of evaluation of theories and presuppositions, while his account of presuppositions explicitly denies the existence of such framework-neutral criteria. Earlier we saw that Brown's account of the fundamentality of presuppositions was hard to square with his view that presuppositions can be shown to be erroneous by the results of research; here we see that his account of objectivity is equally hard to square with the account of presuppositions Brown offers.

4. PRESUPPOSITIONS AND THE PHILOSOPHY OF SCIENCE

The status of presuppositions not only bedevils Brown's account of objectivity, it renders suspect Brown's discussion of the revolution in philosophy of science which has moved us from logical empiricism to the new philosophy of science he espouses.[80] Brown is careful, in his discussions of logical empiricism, to offer his commentary on the failings of that movement not as criticism, but simply as explanation of the development of logical empiricism as it followed the direction of its presuppositions.[81] Nevertheless, Brown suggests that logical empiricism's failure, by its own lights, to solve the problems it set for itself counts as "an important reason for looking seriously at alternative approaches to the philosophy of science."[82] Thus, "solving the problems it sets for itself" counts for Brown as a presupposition-neutral criterion of evaluation for assessing presuppositional frameworks. Here again Brown wants both to deny and assert the existence of presupposition-neutral criteria of evaluation of theories and presuppositions. In addition to this recurrent difficulty concerning the status of presuppositional frameworks, Brown never considers how familiar criteria of theory evaluation get played out as they are applied to theory (or framework) choice in the philosophy of science. Consider, for example, the criterion of explanatory power. Could we defend our choosing the new philosophy of science over logical empiricism on the grounds that the former explains more than the latter? (And would this count as a framework-neutral criterion of evaluation of philosophies-of-science?) Such questions are crucial, if Brown is seriously arguing that change in philosophy of science in akin to change in science. Yet such questions are not raised in his book.

Finally, we must note that other problems pointed out earlier concerning theory change in science according to the new philosophy and epistemology of science are problems for Brown's account for theory change in the philosophy of science, in addition to the problem of the status of presuppositions just discussed. We saw earlier, for example, that Brown's account of rationality is deficient because it fails to distinguish warranted from unwarranted judgments and rational from irrational judgments, and that no analysis is offered of the justification of claims. These failings affect the discussion of theory change in the philosophy of science in two ways. First, most obviously, we have no way of making warranted or rational judgments concerning rival philosophies of science, since we have no way of distinguishing such judgments from unwarranted or irrational ones. Thus theory choice in the philosophy of science can only be arbitrary and irrational. Second, and more importantly, we have at hand framework-neutral criteria by which to judge the new philosophy of science vis-a-vis logical empiricism. Brown, like the logical empiricists, agrees that questions concerning the rationality of judgments, the warrant and justification of claims, and so on are crucial questions that any philosophy of science must grapple with. If the logical empiricist's answers to these questions were unsatisfactory, at least they grappled long and hard with them. The new philosophy of science, on the other hand, deals only cursorily with such questions, and fails seriously to come to grips with such central, if nagging, questions as the nature of the warrant and justification of scientific claims, and the nature of the criteria for distinguishing between rational and irrational judgments, or between rival sets of presuppositions. Thus, not only are there severe problems with Brown's attempt to construe change in philosophy of science as of a piece with change in science; by his own lights it appears that the latest revolution in philosophy of science is unjustified, and that logical empiricism, despite its obvious failings, is superior to the new philosophy of science.

We must conclude, then, that the new philosophy of science, and the new epistemology of science that goes with it, do not succeed. Both fail to confront the deepest questions that need to be confronted. How do we decide among competing sets of propositions? How do we distinguish rational from irrational judgments? What is the nature of the warrant, and justification, of scientific claims? Such questions as these are not squarely faced. The new philosophy of science simply stops its analysis before the most basic questions are reached. We are indebted to Brown for bringing together the diverse strands of post-positivist philosophy and epistemology of science that have been developing over the last two decades. Nevertheless, the new philosophy of science, as presented by Brown, has difficulties at least as severe as the older philosophy of science it seeks to replace. In particular, in the present context, we must note that Brown's failure to present a plausible conception of the new philosophy of science includes a failure to establish a compelling form of relativism.[83]

In this chapter and the last we have considered several attempts to refine, extend, and defend a Kuhnian version of epistemological relativism as part of the general effort to articulate a new philosophy of science.[84] Not all participants in this effort, however, are as kindly disposed to Kuhnian relativism as the authors we have been considering. In the next chapter, we turn our attention to contemporary philosophers who reject Kuhn's general orientation to the philosophy of science, and in particular his relativism, but who nevertheless flirt with a relativistic portrayal of science.[85]

CHAPTER 6

THE UN-KUHNIANS: RELATIVISM VIA THE PROBLEM-SOLVING THEORY OF RATIONALITY

A central problem for the new philosophy of science is that of characterizing the rationality of science, in particular the rationality of theory choice. We have seen in the last three chapters that on a Kuhnian conception of science a satisfactory account of rationality is elusive. There have been, however, non-Kuhnian efforts at articulating science's rationality. In this chapter we consider two such efforts.

Both Stephen Toulmin[1] and Larry Laudan[2] offer extended criticism of Kuhnian philosophy of science,[3] including Kuhnian relativism, in the course of developing their own positive views concerning theory change and the rationality of science. Their independent treatments both characterize science's rationality in terms of *problem-solving* ability: science is rational to the extent that it successfully solves problems. Toulmin and Laudan fill out their views in different ways, and in what follows I attempt to present, and criticize, their respective analyses. I will argue that their approach to rationality in science, in terms of problem solving, suffers from insurmountable difficulties, and cannot be sustained; that rationality in science cannot be construed successfully in terms of problem-solving ability. Especially problematical, I will claim, is the account given of the alleged *truth-independence* of rationality, a key component of the problem-solving approach. In addition, I will argue that, despite their explicit rejection of relativism, both Toulmin and Laudan unwittingly commit themselves to relativistic portrayals of science. This chapter offers, then, both a critique of the problem-solving approach to the rationality of science, and a continuation of the consideration of relativism in the new philosophy of science.

(A) TOULMIN ON RATIONALITY AND PROBLEM SOLVING

For Toulmin, a person's rationality is not reflected by his beliefs, but rather by the way he changes his beliefs:

> A man demonstrates his rationality not by a commitment to fixed ideas, stereotyped procedures, or immutable concepts, but by the manner in which, and the occasions on which, he changes those ideas, procedures, and concepts.[4]

Toulmin focuses his account on changing *concepts*, as opposed to changing beliefs, theories or hypotheses. It is the rationality of the adoption of new concepts, the rejection of old concepts and the maintenance of existing concepts that Toulmin addresses himself to: science is rational to the extent that its adoption, rejection, and maintenance of concepts is rational.[5] How do we establish the rationality of our judgments concerning the warrant of concepts? Not, according to Toulmin, by appealing to a universal impartial rational standpoint – the "Court of Reason" – as philosophers from Socrates to Frege have suggested.[6] The fact of conceptual diversity – the fact that people of different cultural milieus adhere to different, incompatible conceptual schemes – renders the philosophical ideal of an impartial, universal standpoint suspect, since the formulation of such a standpoint always takes place from within some particular cultural/historical setting, and so at best captures that setting's view of impartiality – a view typically not shared by the members of alternative settings. Thus the fact of conceptual diversity threatens to undermine the philosophical ideal of an impartial, universal standard by which competing concepts (claims) can be objectively assessed. Philosophers have tried to get around this problem, according to Toulmin, by establishing "a single, unchanging, and uniquely authoritative system of ideas and beliefs."[7] Such a system was exemplified by logic and geometry, and

> the rational merits of an intellectual position were identified with its logical coherence; . . . the philosopher's measure of a man's rationality become [sic] his ability to recognize, without further argument, the validity of the axioms, formal entailments, and logical necessities on which the claims of the authoritative system depended.[8]

Thus is born what for Toulmin stands as the fundamental mistake of the Western philosophical tradition: the equation of rationality with "logicality." It is this equation of rationality with logicality which creates the seemingly insurmountable problem of conceptual change. For Toulmin argues that there are only two (equally bad) solutions to the problem of the rationality of conceptual innovations, once the alternatives are limited to some species of logicality: absolutism and relativism.

Absolutism, according to Toulmin, seeks to establish universal timeless concepts and standards of rational judgment. History, and the fact of conceptual diversity, are, for the absolutist, strictly speaking irrelevant to the establishment of such concepts and standards. The philosophical task, for the absolutist, is to construct a formal system of timeless concepts and logical relations. Rationality is then simply a matter of judging in accordance with that system. Toulmin argues that this sort of absolutist account of rationality will not do. First, it is doomed to irrelevance for an important class of cases – namely, those cases in which we want to evaluate the rationality of actual conceptual developments. Formal systems, says Toulmin, cannot be helpful in "comparing the merits of alternative concepts within an historical sequence";[9] such systems cannot help us judge the rationality of any actual case of conceptual development in any given discipline. Toulmin argues further that the absolutist approach distracts attention from important "questions about the external application of conceptual systems, as put to practical use."[10] Finally, Toulmin argues that a formal analysis, at best, can only work from *within* a frame of concepts:

> Any attempt to judge conceptual novelties in science, or to make comparisons across the intellectual boundaries between rival systems, soon drives us beyond the range of a purely formal analysis.[11]

If these difficulties doom absolutism, the alternative – relativism – fares no better. Relativism is clearly relevant to actual historical cases of conceptual change – indeed, it is motivated largely by the fact of conceptual diversity – and so avoids the "irrelevance" charge Toulmin levels at absolutism. In recognizing the fact of conceptual diversity, we recognize not only that one's beliefs and presuppositions are "determined" by one's culture, but also that "what kinds of phenomena we regard as puzzling or self-explanatory . . . what types of scientific argument and evidence we find cogent and plausible"[12] are similarly determined. Since "enculturation" of presuppositions and evaluative attitudes runs so deep, the relativist is forced to conclude that any given rational standard can carry weight only within its original cognitive-cultural context. There is, according to the relativist, no rational standard which transcends its particular milieu; and consequently no neutral way to assess the rationality of conceptual change. Relativism is thus self-defeating as an account of the rationality of conceptual change, for it denies the possibility of neutrally evaluating conceptual changes, and so cannot provide an account of the intellectual warrant some changes may have.

Absolutism and relativism are, according to Toulmin, both inadequate as accounts of rationality; they both share the assumption that rationality equals logicality. Toulmin suggests that a successful account of the rational warrant of conceptual change must give up the "rationality equals logicality" assumption, and seek to establish rational warrant in some other way. His suggestion, predictably enough, is that we seek to establish the rational warrant of doctrines by examining "*the conditions on which, and the manner in which [a person] is prepared to criticize and change those doctrines as time goes on*."[13] Rationality for Toulmin is thus essentially dynamic; to judge the rationality of any particular concept (claim) one must judge its evolution or change through time. A discipline or "intellectual enterprise" is rational insofar as it has established rational procedures for change. Such a procedure is rational insofar as it allows the discipline to satisfy further its disciplinary goals.

To explicate his evolutionary account of rationality, Toulmin draws an extended analogy between conceptual change and evolutionary change in biology. Evolutionary theory distinguishes between random variation, by which novel changes in species come about, and selective retention, by which those changes are "selected" for their "fit" in the particular ecological niche in which the organisms find themselves. Analogously, Toulmin argues, we must distinguish between the variations of concepts circulating within a discipline, and those conceptual modifications which are *effective* – that is, which are incorporated into a discipline.[14] Consequently, in accounting for conceptual change, we must separate questions concerning the *innovation* of new concepts (and new variants of old concepts) from questions concerning the *acceptance* of new concepts.

Disciplines consist of bodies or populations of concepts, methods and fundamental aims. They evolve in all three respects. New concepts, however, are the main agents of evolutionary change, and it is in explicating the nature of the evolution of concepts that Toulmin relies most heavily on the Darwinian model. Novel concepts are selectively retained, and become a part of the conceptual population of a discipline, if they fit their intellectual niches. What, then, is an intellectual niche? What demands does a niche place on a concept? How does a concept successfully meet those demands? To answer these questions is to move closer to the heart of Toulmin's account of the rationality of scientific change.

Toulmin argues that disciplines are individuated in terms of *problems*: the problems taken to be central, over time, individuate and unify disciplines.[15] How does we identify problems? "Problems arise . . . where our ideas about the world are at variance either with Nature, or with one another; i.e., where our current ideas fall short . . . of our intellectual ideals."[16] Problems thus depend, for their being problematic, on our intellectual and explanatory ambitions and ideals. So: for Toulmin, disciplines underlie conceptual change; problems underlie disciplines; and intellectual ideals and explanatory ambitions underlie problems. To assess the rationality of conceptual change, then, we must turn our attention to the intellectual and explanatory ideals of a discipline, for only by doing so can we understand the nature of the problems the discipline takes as central, and only by doing that can we assess the "fit" between a novel concept and its intellectual "ecological niche." Toulmin illustrates and fills out this approach with reference to the example of atomic physics. Atomic physics constitutes a discipline in that, since its inception with Thomson and Rutherford, it has held to a more or less constant intellectual ideal – namely, to explain the gross properties of matter in terms of sub-atomic structure:

> . . . the general conception of a 'complete explanation' of the physical properties of matter, given in terms of 'atomic sub-structure,' imposed an intellectual unity on the questions of atomic physics and provided the profession of atomic physicists with its collective goal.[17]

Given the collective ideal of explaining the properties of matter in terms of sub-atomic structure, the discipline of atomic physics can focus on a clearly defined set of problems: namely, those questions that cannot be answered in terms of the concepts and theories accepted in the discipline at a given time:

> . . . scientists locate and specify the intellectual shortcomings of their current concepts by recognizing the shortfall between their current ability to 'account for' the relevant features of the natural world and the explanatory ambitions defined by their current ideals of natural order, or models of complete intelligibility. In short,
> Scientific Problems = Explanatory Ideals – Current Capacities.[18]

So, a problem is a question which a discipline cannot currently answer, but which it ought to be able to answer, according to its intellectual and explanatory ideals. As Toulmin is quick to point out, the problems of a discipline themselves evolve, so that, for example, the problems worried about by Thomson and Rutherford are different from the problems taken seriously by more recent atomic physicists. This is to be expected, since the current capacities of a discipline change over time.

Nevertheless, it is the relative constancy of explanatory ideals that unifies a discipline and licenses our regarding Rutherford and, for example, Weinberg, as members of the same discipline.[19]

Where, finally, does the rationality of conceptual change come in, on Toulmin's account? The selection of a novel concept is rational if and only if it allows the selecting discipline to increase its current capacities towards meeting its explanatory ideals, and thereby contributes to the solution of scientific problems. Thus the rationality of scientific change, for Toulmin, is a function of science's problem-solving ability. Problem solving is in turn related to progress, since solving a disciplinary problem is equivalent to making progress towards reaching a discipline's explanatory and intellectual ideals. So, for Toulmin, scientific change is rational insofar as it contributes to disciplinary progress and problem-solving ability; and consequently, a scientist is irrational when retaining a concept "beyond the point at which the concept loses its explanatory fruitfulness."[20] As an intellectual discipline, a science is rational insofar as it has institutionalized procedures for assessing and criticizing the problem-solving ability of current and novel concepts.[21]

The rationality of maintaining and selecting concepts depends squarely, for Toulmin, on the ability of those concepts to contribute to the solution of disciplinary problems, and consequently, to the furtherance of a discipline's intellectual and explanatory ideals. What of the rationality of those ideals? How can we tell when a discipline's intellectual and explanatory ideals themselves count as rational? Perhaps more importantly, how do we assess the rationality of *change* in disciplinary ideals? If, as Toulmin claims, the selection of a conceptual variation "is justified by showing that it best succeeded in resolving the outstanding problems of a science, and so in increasing its explanatory power,"[22] then such justification (as Toulmin points out) depends on agreement within the science concerning that science's intellectual and explanatory ideals. Without such agreement, there would be no way to determine a concept's ability to increase that science's explanatory power. What happens with respect to rationality, then, when a science's ideals are not generally agreed upon?

Toulmin terms such cases, in which conceptual selection is hampered by a lack of agreement among members of a discipline as to that discipline's intellectual and explanatory ideals, "cloudy". Disputes in cloudy cases cannot be resolved by appeal to intellectual and explanatory ideals, since those ideals are precisely what is in dispute in such cases. Cloudy cases thus constitute a serious threat to the account of rationality in science Toulmin has offered thus far, which hinges on agreement regarding a discipline's ideals. Moreover, appeal to problem-solving ability in such cases is futile, since what problems a discipline seeks to solve, as well as criteria for successful solution, depend on the discipline's explanatory ideals. When ideals are cloudy, so is problem-solving ability, since the latter depends on the former.

Cloudy cases, Toulmin argues, are rationally settled by "appeal, not to the codified rubrics of an established theory, but to broader arguments involving the comparison of alternative intellectual strategies, in the light of historical experience and precedents."[23] The disputants in cloudy cases turn "to considerations of an essentially historical kind"[24] in deciding the future course of their discipline's explanatory ideals; the arguments made in defense of alternative ideals are "common law" arguments, rather than "formal" ones. Rationality in settling cloudy cases, then, is a matter of appeal to historical precedent and careful assessment of the strengths, weaknesses, range and limitations[25] of alternative disciplinary ideals, in an effort to "fulfill most completely [the discipline's] general ideals of scientific understanding."[26]

(B) PROBLEMS WITH THE PROBLEM-SOLVING ACCOUNT OF RATIONALITY

Toulmin's account of rationality in science – in clear cases, rationality is a function of problem-solving ability and progress toward the realization of disciplinary ideals; in cloudy cases it is a function of "common-law" argument which appeals to historical precedent in attempting to extend the evolution of those ideas in a way most compatible with the general ideals of scientific understanding – is unsatisfactory in several respect. While some of the difficulties are unique to Toulmin's analysis, others are more general in that they shed light on other influential efforts to analyse rationality in science, especially that of Larry Laudan. I shall consider the Toulmin-specific difficulties first.[27]

(1) Toulmin consistently conflates "concept" with "proposition framed in terms of some concept", so that his argument throughout confuses claims concerning the rationality of accepting a concept into the family of disciplinary concepts with claims concerning the rationality of accepting some proposition into the body of disciplinary wisdom. Thus we are forced to shift from considerations of rational theory choice to those

of rational concept choice. It may well be that there are important philosophical questions concerning the latter. But such questions are, nevertheless, distinct from questions concerning the former, which are more central epistemologically. Once the distinction between propositions and concepts is noted, much of Toulmin's account decreases in plausibility. For example, consider Toulmin's claim that:

> . . . scientists locate and specify the intellectual shortcomings of their current concepts by recognizing the shortfall between their current ability to 'account for' the relevant features of the natural world and the explanatory ambitions defined by their current ideals of natural order, or models of complete intelligibility. In short,
> Scientific Problems = Explanatory Ideals – Current Capacities.[28]

In the first statement, it is the shortcomings of *concepts* that are identified by their failure to solve current disciplinary problems. In the equation that follows, however, it is the shortcomings of a discipline's *current explanatory capacities* which are so identified. This passage clearly suggests, indeed assumes, that a discipline's explanatory capacity is solely a function of its stock of concepts. But this is patently false: explanatory power is a function, not simply of concepts, but of what is done with – that is, what claims are made in terms of – those concepts. Explanatory power, that is, is a property of propositions couched in terms of concepts, not of concepts *simpliciter*. A simple example makes this clear. Consider the discovery of irregularities in the orbit of Uranus, during the heyday of Newtonian physics. The Newtonian programme was, at this point, well developed, with a rich stock of concepts. Nevertheless, Uranus' orbit could not be explained satisfactorily, given the explanatory ideals current at the time. As is well known, Leverrier and Adams sought to explain the irregularities by postulating a new planet, and after calculating (with the existing stock of concepts and propositions of Newtonian theory) the location and mass the new planet must have in order to account for Uranus' eccentricities, Neptune was found. Here is a clear case of increased explanatory power: before Neptune's discovery, Uranus' orbit could not be "accounted for"; after the discovery it could. But the new ability to explain Uranus' irregular orbit was not a result of new concepts, nor a shift in "ideals of natural order." It was the result, rather, of conceiving, formulating and verifying the proposition that there was a hitherto unknown object in space causing Uranus' orbit to deviate from its predicted path. If we but note that the work of Adams and Leverrier required the addition of no new concepts, but was carried out entirely in terms of the stock of existing concepts (mass, planet, gravitational force, orbit, space, etc.), we can see the difficulty with Toulmin's conflation of concepts and claims, especially if they are related (as Toulmin argues) to explanatory power. The rationality of science cannot be understood solely, or even mainly, in terms of concepts.[29]

(2) Likewise flawed is Toulmin's discussion of absolutist and relativist approaches to the nature of rationality. Toulmin's criticism of absolutism, recall, is that it avoids the problem of relativism at the cost of irrelevance, in that absolute, formal systems of concepts could not aid our assessment of the rationality of intellectual changes. That is, Toulmin argues that an absolutist stance prevents us from "comparing the merits of alternative concepts within an historical sequence."[30] There are several points to be made against this rejection of absolutism. First, it is false that absolute standards cannot be used to compare the merits of sequential theoretical developments, since these developments can be compared with the absolute standard, and evaluated according to their respective abilities to meet the standard. If absolute standards are understood as regulative and prescriptive, then actual scientific practice and argument can indeed be assessed by appeal to and comparison with absolute standards.[31] Indeed, the possibility of "comparing the merits of alternative[s]" depends on there being some alternative-neutral conception of merit, which itself must count as an absolute standard. Thus, any account of rationality which is not absolutist at least in the sense of having regulative standards of merit is doomed, not to irrelevance, but to impotence – this minimal sort of absolutism is necessary if evaluation is to be even possible. Moreover, Toulmin's approach to rationality, in which the selection of a concept by a discipline is rational if and only if it furthers the ability of the discipline to achieve its disciplinary goals, is itself an absolute standard. Thus Toulmin adopts an absolutist account of rationality of exactly the sort he argues is doomed to irrelevance.[32]

(3) More deeply, Toulmin's analysis of absolutism suffers from defects inherited from his argument that rationality cannot be understood in logical or formal terms. Toulmin argues that formal analysis can work only from within a system of concepts:

> Any attempt to judge conceptual novelties in science, or to make comparisons across the intellectual boundaries between rival theories, soon drives us beyond the range of a purely formal analysis.[33]

There are several difficulties with this contention. For one thing, it is a mistake to think that diachronic events and processes cannot be conceptualized or assessed in formal or logical terms. We can, for example, compare rival theories' ability to make successful novel predictions, their fertility, and their ability to respond to anomaly, in formal terms (e.g., in terms of probability). This is especially easy to see if "logic" and "formal" are taken to refer, not simply to deduction as it manifests itself in the predicate calculus, but more generally to include criteria typically used by scientists to evaluate theories and by philosophers considering the nature of such evaluation. Criteria such as fertility (or progressiveness), predictive power, explanatory power, simplicity, and the like, while not narrowly deductive (though even in these, deduction plays *some* role), are nevertheless logical or formal in that they speak to the relationship between hypothesis and evidence, and moreover function as means of evaluation of hypotheses and theories.[34] Temporal developments can indeed be assessed logically.

Moreover, even if logic is taken not broadly, as above, but more narrowly in terms of deduction, Toulmin's claim still fails. For strict deductive relationships can hold across rival conceptual systems. Consider the propositions

(1) Pa
(2) Pa v Fb
(3) Fb → Fb

These propositions are framed with different sets of concepts, and so can be construed as constituting alternative conceptual systems. Thus, according to Toulmin, strict logical relations cannot exist between them. But this is straightforwardly false, since deductive relations do: (1) logically implies (2) as well as (3), and (2) implies (3). Indeed, even incompatible propositions such as

(4) Ra.Sb.Tc
(5) Ra.~Sb.Qc

can be formally related, in this case by the relation "__ shares a common deductive consequence with __", namely ⌜Ra⌝. So Toulmin is mistaken in his claim that formal relations cannot obtain across rival conceptual systems.[35]

(4) It is worth noting that Toulmin's view that rationality is a matter of the conditions under which a person changes his point of view does not,

of itself, constitute a rival to the view that rationality is a function of logicality. Toulmin's insistence that we construe rationality in terms of conditions for change of belief, or of one's manner of holding beliefs, only puts the logicality question back one step. For we have no reason to think that those conditions or manners cannot themselves be put in logical terms – the more so since, on Toulmin's view, those conditions and manners are rational insofar as they take properly into account explanatory ideals and explanatory power. Thus Toulmin's positive account of rationality, even if accepted, does not necessarily conflict with the account he opposes, namely that rationality is a function of logicality.

(5) Toulmin's discussion of rationality in "intrinsically cloudy" cases faces severe difficulties. Recall that cloudy cases are those cases in which a discipline's explanatory ideals are themselves in dispute. In such a case, conceptual/theoretical novelties cannot be rationally assessed in terms of their ability to meet those ideals, and so to further disciplinary goals, as they are (according to Toulmin) in "clear" cases. Moreover, problem-solving ability likewise cannot be appealed to, since what counts as a problem depends on explanatory ideals. Rather, in cloudy cases novelties are assessed in terms of historical precedent by way of "common law" arguments and considerations, the aim being to choose that alternative explanatory ideal which will "fulfill most completely [the discipline's] general ideals of scientific understanding."[36] Toulmin offers several examples of such cloudy cases, one of which is the recent history of micro-physics, in which some scientists, most notably Geoffrey Chew, argued that microphysics should give up the explanatory ideal of explaining gross phenomena in terms of atomic sub-structure.[37] One obviously cannot appeal to the reigning explanatory ideal of microphysics to settle this dispute between Chew and more orthodox physicists, since it is that ideal which is itself being called into question. Toulmin argues that such cases are to be settled, not by appeal to disciplinary ideals, but to "broader arguments involving the comparison of alternative intellectual strategies, in the light of historical experience and precedents."[38] He illustrates his solution in terms of historical precedents by witnessing subsequent debate in the particle-physics case. Chew and those in favor of changing the explanatory ideals of microphysics appealed to the precedent of Duhem wrongly rejecting Thomson's suggestions concerning explaining gross phenomena in terms of subatomic structure. Orthodox theorists countered by citing the precedent

of the rebel Prout, who speculated prematurely in challenging Dalton's atomic theory. Thus each side sought historical cases which they could analogize to the present situation, each choosing so as to make their own side "look good". And, as Toulmin summarizes the process, "so the search went on, for the most exact and revealing historical precedents."[39]

Why should such search for historical precedent serve to guide rational choice in cloudy cases? It is not clear that it should. Suppose, for example, that both the rebels and the orthodox theorists agreed that the present case was similar in many respects to the Prout episode. Would this settle the debate? Of course not. For the rebels could argue that Prout was mistaken in that case, but in the present case the rebel stance articulated by Chew is correct, so that a disciplinary "change of stream" is called for. After all, were all to agree as to the proper interpretation of the Prout case – that Prout's challenge to Dalton was indeed misguided, rather than that Prout's challenge was legitimate and that micro-physics went wrong at that point – then all would thereby be agreeing that the explanatory ideal which favoured Dalton was correct, or at least appropriate, for judging that dispute. Thus the judgment of rebel and orthodox alike would be based on a common view of the proper disciplinary ideals for micro-physics at that time. But such shared opinion is precisely what is lacking in cloudy cases, as the rebels argue that the historical course should no longer be followed. So it is not clear that agreement about historical precedents would aid at all in rationally settling current cloudy disputes.

Toulmin argues that appeal to historical precedent would contribute to the rational resolution of cloudy dispute, by helping the two sides to estimate "the most promising lines for future theoretical development."[40] Here Toulmin draws an analogy with certain legal situations, for example, Supreme Court cases, in which an appeal to precedent is unhelpful. In such a case, the court must *extend* the law, and *make* precedent, by re-analyzing "the social functions of the law in its application to some novel historical situation."[41] Such novel judgments, Toulmin argues, are judgments concerning "the way in which the law should develop, at this point in historical time, in order to fulfill most completely its general ideals of equity, humanity, and certainty."[42] But such appeal to general ideals, which are agreed upon by all disputants in such judgments, destroys the analogy Toulmin tries to draw, for, *ex hypothesi*, such ideals do not exist in cloudy cases – if they did, the case would not be a cloudy one. The general ideals of the law are analogous to the explanatory ideals of a discipline, in Toulmin's analogy, and these latter ideals are precisely what cannot be appealed to in cloudy cases. And, since a genuine extension of such ideals is necessary, it is not clear that appeal to prior ideals, even if possible, would help. For the prior ideals could not serve to select from among novel alternatives – if they could, the selected alternative would constitute a choice *based on* those prior ideals, not an *extension* of them – just as, in the legal case, prior precedent cannot serve to guide genuinely precedent-*making* decisions, for, if it does so guide, the new decision is precedent-*following*, not precedent-*setting*. Thus there is a clear sense in which the appeal to precedent is impotent, and the appeal to history useless, in extending prior ideals.[43]

Moreover, Toulmin's claim that such (legal precedent-setting) judgments are rational[44] is gratuitous. In such cases as Toulmin describes, the court's decision is deemed by him to be rational, but the only constraint on such decision is that the court use its judgment. Thus, any decision based on judgment is rational. Now, judgment may well be a necessary condition of rational judgment, but it is surely not a sufficient one. To analyze rationality in this way is to give up the very possibility of irrational judgment, since any judgment counts on this analysis as rational. The Supreme Court Justice who fails to eliminate bias, reasons poorly, misuses evidence, or fails to take into account probable consequences of her judgment judges rationally, on Toulmin's analysis, so long as she judges, perhaps erroneously, that the decision so reached furthers the general ideals of the law (which may well be in dispute). It is not all clear that such a judgment should count as rational, but it must be counted as rational on Toulmin's analysis.

Finally, the analogy that Toulmin draws between such legal judgments and scientific judgments in cloudy cases leaves the rationality of the latter in circumstances as desperate as those of the former. For here, too, a discipline's "general ideals of scientific understanding" are what are at issue, and so Toulmin's appeal to them[45] is both impossible (*ex hypothesi*) and question-begging. And, as before, Toulmin's analysis of rationality comes simply to this: judgment must be used. This seems to me as unsuccessful here as it was a paragraph ago. For the several reasons discussed, then, it must be concluded that Toulmin's analysis of the rationality of scientific choice in cloudy cases is unsuccessful.

(6) As argued earlier, Toulmin's account of rationality in clear cases

(i.e., in cases where disciplinary ideals are not in dispute) seems to force him into an absolutist position, in that rational choice in clear cases is simply a matter of choosing so as to fulfill disciplinary ideals – this latter being an absolute standard. This absolutism is something of an embarrassment for Toulmin, in that he argues against absolutism and claims to be offering a non-absolutist account.[46] If Toulmin's analysis of clear cases drives him, against his will, to absolutism, his analysis of cloudy cases offers an opposite peril – namely, it commits Toulmin to relativism. For if the argument advanced above is correct, Toulmin's analysis of cloudy cases leaves choice in such cases relativistic, in that the only constraint on the chooser is that she use her judgment. Choosers who espouse rival disciplinary ideals, and so rival views as to the importance of particular problems, and who therefore judge oppositely, have no more stringent constraints to appeal to, and, to the extent that their judgment is unconstrained, choose relative to their predilections. Thus, while Toulmin argues against both absolutism and relativism, his analysis of scientific choice in clear cases in absolutistic, and in cloudy cases, relativistic.[47]

(C) LAUDAN, PROBLEM SOLVING AND THE TRUTH-INDEPENDENCE OF RATIONALITY

We have seen several difficulties with Toulmin's view of the rationality of scientific choice. There are other difficulties to note. However, these others are, I believe, more general than those discussed above in that they are difficulties, not just with the specific account Toulmin offers, but with any view which analyses the rationality of science in terms of problem-solving ability, or that takes as fundamental that science is a problem-solving activity. Another major effort to analyze the rationality of science along these lines is Larry Laudan's *Progress and Its Problems*. Before moving on to the general discussion of difficulties with the "problems" approach to rationality in science, then, I will first (briefly) present Laudan's approach, and highlight the similarities between Laudan's view of the rationality of science, and Toulmin's.[48]

Laudan takes as fundamental that "science is essentially a problem-solving activity," and that "the aim of science is problem-solving."[49] Progressiveness in science is a function of problem-solving effectiveness:

> . . . *any time we modify a theory or replace it by another theory, that change is progressive if and only if the later version is a more effective problem solver . . . than its predecessor.*[50]

and rationality is a function of progressiveness:

> *To make rational choices is, on this view, to make choices which are progressive* (i.e., which increase the problem-solving effectiveness of the theories we accept).[51]

Thus, for Laudan, rationality in science is a function of problem-solving effectiveness: we choose rationally if and only if our choices maximize our ability to solve problems, and so further our scientific aims. Laudan, like Toulmin, casts rationality as a function of problem-solving effectiveness (the latter, recall, argues that the choice of a concept is rational if its choice helps further disciplinary goals, by solving problems); like Toulmin, Laudan eschews the traditional link between rationality and truth:

> . . . it has normally been held that any assessment of either rationality or scientific progress is inevitably bound up with the question of the *truth* of scientific theories. Rationality, it is usually argued, amounts to accepting those statements about the world which we have good reason for believing to be true. Progress, in its turn, is usually seen as a successive attainment of the truth by a process of approximation and self-correction. I want to turn the usual view on its head by making rationality parasitic upon progressiveness. *To make rational choices is, on this view, to make choices which are progressive* (i.e., which increase the problem-solving effectiveness of the theories we accept). By thus linking rationality to progressiveness, I am suggesting that we can have a theory of rationality *without presupposing anything about the veracity or verisimilitude of the theories* we judge to be rational or irrational.[52]

So, in their own ways and in their own vocabularies, and despite important differences in their accounts, both Toulmin and Laudan argue that problem solving is the basic task of science (for Toulmin, this is couched in terms of furthering disciplinary and explanatory ideals); and both analyze the rationality of science in terms of progressiveness and problem-solving effectiveness: rationality is a function of problem-solving effectiveness; that effectiveness, and attendant progressiveness, are primary. This general move, of analyzing rationality in terms of problem-solving effectiveness, suffers from major difficulties – so grave, I believe, that the approach is unlikely to yield a satisfactory account of the rationality of science.

(1) First, Laudan's account suffers from a difficulty similar to the difficulty discussed above concerning Toulmin's "cloudy" cases. The problem with cloudy cases, recall, is that, when there is no agreement within a discipline concerning disciplinary and explanatory ideals, there can be no agreement as to what constitutes a satisfactory solution to a problem, or even as to what counts as a problem, since those things are

determined by reference to disciplinary and explanatory ideals. As argued above, Toulmin's appeal to historical precedent fails to secure rationality for decisions made in such "cloudy" cases. For Laudan, a similar difficulty arises: if rationality is a function of problem-solving effectiveness, how is rationality determined in cases in which there is no general agreement, among rival research traditions, concerning what counts as a problem and as a solution? Laudan's complex and insightful discussion of problems deals almost exclusively in terms of problems *within* traditions – indeed, it seems clear that for Laudan, as McMullin notes, "what counts as a problem is determined ultimately by the research tradition."[53] Laudan's only discussion of problems and problem-solving effectiveness, and so rationality, *across* research traditions, are his suggestions that (1) problems *can* be recognized, by rival, incommensurable research traditions, as the same problem; and (2) that even if (1) were false, rival research traditions could be assessed via internal examination – that is, by asking, for any given research tradition, whether it "has solved the problems which it set for inself".[54] By such internal evaluation, "we can come up with a characterization of the progressiveness . . . of the research tradition.

If we did this for all the major research traditions in science, then we should be able to construct something like a progressive ranking of all research traditions at a given time. It is thus possible, at least in principle and perhaps eventually in practice, to be able to compare the progressiveness of different research traditions, *even if those research traditions are utterly incommensurable in terms of the substantive claims they make about the world*![55]

Unfortunately, neither of these ways of assessing the rationality of rival research traditions will work. (1) is right as far as it goes, but that is not nearly far enough. For while Laudan is surely correct that radical incommensurability claims are false and that some problems can be recognized by rival research traditions as the same problem, it nevertheless remains the case that proponents of rival traditions often assess the nature and the importance of such problems very differently, and so, while agreeing that P is a problem, would greatly disagree about the relative importance of P in assessing the merits of rival research traditions. For example, consider the rivalry between pre-Daltonian chemistry and Daltonian chemistry, and the problem of explaining various qualitative changes substances undergo during chemical reactions. Both traditions recognized this as a problem. But the former tradition took the problem to be a serious one, and their ability to solve it a major point in favor of their tradition, while the latter tradition took the problem to be a minor one, and viewed it as hardly a problem at all that their tradition could not solve it. Thus the fact that both traditions recognize *P* as a problem (and so, in general, that Laudan is right that "with respect to any two research traditions . . . in any field of science, these [sic] *are some joint problems which can be formulated so as to presuppose nothing which is syntactically dependent upon the specific research traditions being compared*,")[56] does little to ensure the success of efforts of appraisals of progressiveness and rationality across rival research traditions. For, even though rival traditions might agree that *P* is a problem, they may still disagree wildly as to the importance of *P*, and so as to the weight to be attached to the success (failure) of solving (not solving) *P*. Since what counts as a problem, and its importance, is for Laudan ultimately a function of the research tradition, and since such assessments will vary from tradition to tradition, the ability of rival traditions to identify common problems is only a meager first step in an argument purporting to show that appraisal of progressiveness and rationality across traditions is possible. For the argument to succeed – that is, for it to be shown that such cross-tradition appraisal is possible – one must accomplish the much harder tasks of showing (1) how rival traditions can agree as to what counts as a successful solution of a common problem, and (2) how agreements as to the importance of solving (not solving) such problems can be had. Laudan's first suggestions does neither of these, and so does not at all show how appraisal of progressiveness and rationality across rival research traditions is possible.[57]

Laudan's second suggestion, that rationality across rival research traditions can be assessed by assessing progressiveness internally – that is, by determining the success of a tradition in solving the problems that it generates or recognizes for itself – fares no better than his first suggestion. For one thing, a tradition may wrongly inflate its ratio by failing to recognize problems it ought to come to grips with, and so fail to "set for itself" problems it should set for itself. Moreover, by setting its sights conservatively, a cautious tradition may secure for itself a higher assessment of progressiveness than a more bold tradition which suffers failures to solve problems because it does not hesitate to set for itself difficult problems. Thus a bold tradition might have a relatively low rate of progressiveness, despite its success at solving difficult problems, because of its failure to solve other difficult problems it sets for itself; while a less bold tradition can secure for itself a relatively higher rate of progressiveness simply by avoiding such problems. In addition, such internal assessment would make it impossible to distinguish the relative progressiveness of rival theories (e.g. Newton vs. Einstein) from that of non-rivals (e.g. Freud and Einstein).[58] Most importantly, the problem discussed above concerning Laudan's first suggestion resurfaces here: since the measure of internal progressiveness would need to take into account, not just the *number* of problems solved relative to the number recognized, but also the *weight* and *importance* of those problems, the internal measure would be objectionably internal in that its calculations and computations of progressiveness would be disputed by rival traditions which weighted problems differently from the tradition whose progressiveness is being calculated.[59] Thus here too, as earlier, Laudan has failed to show how appraisal of rationality and progressiveness across rival research traditions is possible. As with Toulmin's analysis of cloudy cases, Laudan's account of assessing problem-solving effectiveness, and so rationality, across rival research traditions is more relativistic than he supposes.[60] Unless a non-tradition-bound criterion for establishing what counts as a problem, and for establishing the relative importance of problems, can be had, it is difficult to see how such relativism can be avoided. And it is imperative to note that problem-solving effectiveness cannot constitute such a criterion, since problem-solving effectiveness is itself determined by appeal to particular research traditions, and will vary from tradition to tradition.[61]

(2) Second, both Toulmin and Laudan flounder in their respective discussions of how the analysis of rationality in terms of problem-solving ability relates to the role of truth, both as an aim of scientific inquiry, and as a notion conceptually related to rationality. Their particular analyses of the role of truth fail; in addition, their enterprises – to provide "truth-independent" accounts of the rationality of science – turn out, on the accounts of rationality they offer, to be incoherent in that those accounts appeal to truth in order to establish truth-independence.

Toulmin denies that scientific theories or claims say anything true or false about the world; rather, he holds that claims and theories assert applicability. On this line to say, for example, that "light travels in straight lines" is not to commit oneself to the truth of that assertion, but rather to the successful applicability of certain of our theoretical representations and techniques to light beams.[62] But as Kordig points out, on the standard Aristotle-Tarski conception of truth, to say that a claim can be applied *successfully* to an aspect of the world is simply to say it is true of that aspect of the world.[63] The upshot of successful application of explanatory or representation techniques simply is truth.[64]

Laudan's attempt to sever the conceptual link between rationality and truth, and to develop a truth-independent account of rationality, is more systematic and serious than Toulmin's. Nevertheless, it is flawed. First, Laudan's argument for the truth-independence of rationality is unsuccessful; second, despite his argument for truth-independence, Laudan's positive account of rationality in fact appeals (as several writers have noted) to considerations of truth at crucial junctures. I will develop these points in turn.

As noted earlier, Laudan claims that by "linking rationality to progressiveness, I am suggesting that we can have a theory of rationality *without presupposing anything about the veracity or verisimilitude of the theories* we judge to be rational or irrational."[65] The reader might well wonder why we would want such a truth-independent account of rationality. Laudan's answer is straightforward: we want assessments of rationality to be truth-independent, because accounts of rationality which link rationality with truth are doomed to failure:

> Philosophers and scientists since the time of Parmenides and Plato have been seeking to justify science as a truth-seeking enterprise. Without exception, these efforts have floundered because no one has been able to demonstrate that a system like science, with the methods it has at its disposal, can be guaranteed to reach the "Truth", either in the short or in the long run. *If rationality consists in believing only what we can reasonably presume to be true, and if we define 'truth' in its classical, non-pragmatic sense, then science is (and will forever remain) irrational.*[66]

There is, underlying this passage, a subtle confusion which, once brought to the surface, does much to undermine Laudan's motivation for a truth-independent account of rationality. As Laudan notes, for millenia we have tried to guarantee the truth-directedness of science and scientific method. Laudan argues that we have failed – which is to say, we can offer no guarantee that science leads to (even approximate) truth. Thus we cannot say that science is rational in that it secures for our beliefs the status of truth. Truth cannot, that is to say, serve as a sufficient condition for rationality – we cannot say, "it is rational to believe *p*, because *p* is true," for, if *p* is a typical scientific claim, arrived at by the use of scientific method, we have no basis for the clause "because *p* is true." This is the thrust of Laudan's claim that, if we seek to incorporate truth into our account of rationality, that account is doomed to failure – it is doomed to failure because we seem to be unable to secure for scientific beliefs the status of (even approximate) truth.

However, there is a philosophical confusion of cart and horse here. Laudan takes the classical connection between truth and rationality to be an evidential one: *p* is true, and that fact (that *p* is true) counts as grounds for taking our belief in *p* to be rational. Truth, on this line, counts as grounds for rationality. This, I believe, is a mistaken explication of the classical connection between rationality and truth, according to which rationality amounts to believing what we have good reason to believe. Such good reason for believing *p* provides grounds for believing that *p* is (at least somewhat likely to be) true. Rational belief in *p* thus is not grounded or evidentially based on *p*'s being true; rather, the evidential relation is reversed: judgments concerning *p*'s truth are grounded on *p*'s being rationally believed. To say, in short, that we have good reason to believe *p*, is to say, in effect, that we have good reason to believe that *p* is true. We could put this point another way, namely, that the *import* or *upshot* of rational belief is truth. That is, the import of the fact that we have good reason for believing *p* is that *p* is (at least somewhat likely to be) true.[67] We cannot, as Laudan makes clear, appeal to the truth of *p* in order to establish the rationality of believing *p*. But we can, as Laudan agrees, believe claims rationally in that we have good reasons for doing so. Such good reasons – the hallmark of rationality – provide grounds for truth. Their import is truth.

Laudan's argument thus fails to establish the failure of truth-connected accounts of rationality. At best it shows the impossibility of using truth as an evidential criterion for establishing rational belief. But neither we, nor the classical view, need construe the relation between truth and rationality in this way. A better way to conceive that relation is to reverse it: begin, as Laudan does, by recognizing that believing *p* rationally is believing *p* for good reason; then note that such good reasons provide grounds for taking *p* to be true. For the import of claiming that we have *good* reason for believing *p* is that such reason constitutes, as it constitutes good reason for believing *p*, good reason for believing that *p* is true. Here is the classical connection between truth and rationality. On the classical view, truth does not constitute grounds for rational belief; rather, rational belief (good reason) constitutes grounds for truth. Laudan's argument may be effective against the view that truth can play an evidential role in the establishment of the rationality of belief. But since that is not the correct way to construe the relation between rationality and truth, his argument does not in the least establish either the desirability or the necessity of settling for a truth-independent account of rationality.[68, 69]

I have been arguing that Laudan's case for the necessity and desirability of constructing a truth-independent account of rationality is weak, primarily because he misconceives the classical connection between truth and rationality. There is an additional difficulty here. Laudan not only misconstrues the way in which truth and rationality are linked; he also, in his attempt to construct a truth-independent account of rationality, smuggles in considerations of truth in such a way as to render his account of rationality truth-dependent after all. Because several writers have noted this, I will argue the point briefly. Laudan constructs his account of rationality not in terms of truth but in terms of problem-solving effectiveness and progressiveness. Nevertheless, in various respects that account implicitly appeals to truth. For example, a research tradition's problem-solving effectiveness depends in part on isolating (and solving) genuine empirical problems, but what counts as a genuine (and not a pseudo-) problem depends on considerations of truth. That is, assessing problem-solving effectiveness depends on our prior ability to distinguish genuine from pseudo-problems, which in turn depends on "considerations of truth and falsehood."[70] Similarly, problem-solving effectiveness is a function of a theory's ability to avoid refuting anomalies, and to avoid conceptual problems such as inconsistency. Here, too, it seems impossible to conceptualize such factors as anomalies and inconsistencies except by linking them with truth: refuting anomalies and inconsistencies figure badly for a theory's problem-solving effectiveness because they suggest that afflicted theories are not true.[71] A research tradition is marked by its ontology; but here too the significance of Laudan's claim that ontology is an important feature of research traditions is unintelligible except insofar as that ontological element is related to the veracity of the tradition in question.[72] Laudan's general analysis of rationality in terms of the aims of science also appeals to considerations of truth. Science is rational if it attains its end, Laudan argues. But those ends have also to be rational if the enterprise is to be so; and Laudan offers little argument for taking problem solving to be an inherently rational aim.[73] He argues that we justify our scientific activity in terms of our sense of curiosity about ourselves and the world,[74] but such curiosity can only be curiosity about the *truth* concerning us and the world.[75] Despite Laudan's arguments for a truth-independent account of rationality, his own account actually appeals to considerations of truth in the several ways just noted. Considerations of truth are thus built into his allegedly truth-independent account of rationality.[76]

(D) CONCLUDING CONSIDERATIONS

I have been arguing that there are three serious difficulties with Laudan's account of rationality: first, that it cannot adequately or non-relativistically account for determinations of rationality across rival research traditions; second, that is misconceives the classical relationship between rationality and truth, mistakenly argues for the necessity of a truth-independent account of rationality, and wrongly extols the desirability of such an account; and third, despite argument to the contrary, that it is truth-dependent in that it implicitly appeals to considerations of truth at several key junctures. There is yet another problem which any attempt to understand the rationality of science in terms of problem-solving ability and progressiveness must inevitably confront. Suppose an account such as Laudan's proves functionally successful, so that we can more or less straightforwardly pick out theories and research traditions and establish their problem-solving effectiveness. We are then in a position to say, for example, that it is rational for us to accept RT_1, and not RT_2, because RT_1 is a more effective problem solver. Since on Laudan's view rationality is not related to truth, such rational acceptance says nothing about the truth of RT_1. When we say it is rational to accept RT_1, we commit ourselves not to the truth of RT_1, but only to its problem-solving effectiveness.

How are we to understand such effectiveness? How can we as meta-scientists (philosophers of science) explain the fact that RT_1 has greater effectiveness than RT_2? In order to explain RT_1's effectiveness – to explain why it, for example, but not RT_2, could successfully solve problem P – we need to appeal to considerations beyond that effectiveness. So, let us distinguish between two sorts of explanation: first-order explanation of empirical and conceptual problems, and second-order explanation of a research tradition's ability to provide first-order explanations (i.e. solve problems). At best Laudan has shown that first-order explanations which do not appeal to considerations of truth are possible. He has not raised any questions about the nature of second-order explanations. Yet if we are to understand science, we surely need to understand the *significance* of first-order explanations, which includes at least the possibility of second-order explanations. What is the significance of first-order explanations? How we can explain the fact that RT_1 effectively solves P?

One standard way of explaining RT_1's problem-solving ability is in terms of truth. RT_1's successful problem solving is explained in terms of its being right about the world. According to the view now under consideration, if RT_1 is not somehow right about the world, its effectiveness is largely inexplicable. On this view, then, truth functions as an explanatory factor in our overall theory of science.[77]

The "truth explains the problem-solving ability of science" view just mentioned needs considerable explication and defense. I shall not endeavor to provide such defense here. I want simply to point out that a truth-independent account of the rationality and progressiveness of science must confront the problem of second-order explanation. Neither Laudan's nor Toulmin's account does. Laudan, in particular, never raises the possibility of asking for second-order explanations – that is, of explaining how it is that research traditions solve problems. He is content to rest our understanding of science on its first-order abilities. But if we are to have any epistemic understanding of science, we must do more than simply measure problem-solving effectiveness. We must also ask about the significance of such effectiveness; we must explain the fact that RT_1, but not RT_2, can solve P. The view of rationality that Laudan rejects offers such an explanation, by appeal to truth. Laudan has rejected this explanation, but has not offered an alternative. He has simply not addressed this problem, but it is a problem which is raised in an acute form as soon as the explanation in terms of truth is rejected.

So, Laudan's and Toulmin's truth-independent accounts of the rationality of science are flawed. Both of their views face the particular difficulties noted earlier. Especially problematic, given their desire to offer truth-independent accounts, is that both these accounts appeal to truth at critical junctures. They also leave unexplained phenomena – notably, the problem-solving ability they make so much of – that are usually explained by appeal to truth. In short, they do not provide serious alternatives to standard accounts of the rationality of science. Finally, of special interest in the present context is the fact that they fail to overcome the new philosophy of science's predilection for relativism. Despite their explicit disavowals of relativism, both Toulmin and Laudan present accounts of science which are relativistic in the ways noted above. But, because of the problems those accounts face, they fail to support a relativist construal of science.

In Part II we have been concerned with relativism as it arises in the new philosophy of science, and with the latter more generally. However, while the new philosophy of science is one of the major breeding grounds of relativism, it is by no means the only one. In Part III we will turn our attention to some more general epistemological considerations which are held by some to support relativism; we will also consider the prospects for an "absolutist" alternative to relativism in epistemology.

PART III

FURTHER EPISTEMOLOGICAL CONSIDERATIONS

CHAPTER 7

GOODMANIAN RELATIVISM

Nelson Goodman's work is universally regarded as pioneering and fundamental, and his attempts to clarify the nature of induction, symbol systems, art, theorizing and understanding have received and continue to receive great attention. Central to that work is a view Goodman describes as "radically relativist." Goodman's unusual brand of relativism will be the focus of the present chapter. I will first briefly review the development of Goodmanian relativism through his earlier books and articles. Then I will develop a more detailed and elaborate account of the relativist position Goodman develops in his recent *Ways of Worldmaking*. I will argue that there is a crucial ambiguity in Goodman's portrayal and defense of relativism, having to do with the role of "criteria of rightness" in Goodman's system, and that this ambiguity licenses two quite distinct versions of Goodmanian relativism. I will argue that the weak version of relativism is defensible, and important, but that the strong, or "radical," relativist position suffers from a fatal flaw, and fails to avoid the anti-relativist arguments presented in earlier chapters.

A BRIEF HISTORY OF THE DEVELOPMENT OF GOODMANIAN RELATIVISM

Goodman reassures us in the foreword to the latest edition of *The Structure of Appearance* that in his view that structure is not unique; rather, he says, in his characterization of appearance "relativism runs throughout."[1] The relativism that runs throughout *SA* is a *systematizational* one – the view, as one interpreter of Goodman puts it, that "conflicting systematizations can be found for any prephilosophical subject matter."[2] There is, according to Goodman, no privileged system which has any special priority – epistemological or otherwise – in the task of describing reality. Rather, there may be many purposes for constructing such descriptions, and a system that suits one purpose may well be inferior to another system with respect to some other purpose. Moreover, questions of ontology are relative to system. That is, what is "fundamental" in one system need not be in another. Since for Goodman no system is generally superior to every other, and no description can be had save by the use of some system, no claims regarding basic ontology can be made in a systemless or system-neutral way. Thus the pursuit of a basic ontology of the world is for Goodman a pointless endeavor.[3] Choice of system is relative to pragmatic considerations. So whether or not one should construe the world as being made up fundamentally of phenomenal elements such as qualia, or of physical elements such as objects, depends on the purposes of the construal. There is no system-neutral answer to the question "what is there?"[4]

It is important to note that Goodmanian relativism in *SA* does not come to the view that all systems are equally worthwhile, that there is no choosing from among alternative systems. Given pragmatic constraints and specific purposes, it is quite straightforward to establish that, relative to those constraints and purposes, one system is preferable to another. Thus if one wants a characterization of a theory that does not contain "theoretical" predicates, but does not mind an infinity of axioms, one can characterize the theory by the Craig method; whereas if one's purposes are different, so that one is willing to abide theoretical predicates but not an infinite number of axioms, then the Craig systematization is less acceptable than some other. So, while there is no unique, system-neutral answer to the question "does theory T have a finite number of axioms?", it is possible for pragmatic considerations to function as trans-system criteria by which objective evaluations of the worth of alternative systematizations can, relative to those considerations, be made. We shall return to this point below.

Goodmanian relativism is developed further in his essay, "The Way the World Is."[5] Here Goodman makes plain his view that there is no one way the world, or even some small part of the world, is, but rather that there are as many ways the world is as there are true descriptions of it:

> There are very many different equally true descriptions of the world, and their truth is the only standard of their faithfulness. And when we say of them that they all involve conventionalizations, we are saying that no one of these different descriptions is *exclusively* true, since the others are also true. None of them tells us *the* way the world is, but each of them tells us *a* way the world is.

> If I were asked what is *the food* for men, I should have to answer "none." For there are many foods. And If I am asked what is the way the world is, I must likewise answer, "none." For the world is many ways. The mystic holds that there is some way the world is

and that this way is not captured by any description. For me, there is no way that is the way the world is; and so of course no description can capture it. But there are many ways the world is, and every true description captures one of them. The difference between my friend [the mystic] and me is, in sum, the enormous difference between absolutism and relativism.[6]

Here Goodmanian relativism again underscores the point that alternative systematizations are possible, and that there is no "privileged" absolutist framework which better serves to organize our descriptions than other frameworks. If a description is true, it describes a way the world is, even though it is made relative to some system or framework.[7] Relativity of framework thus does not preclude objective characterization of the world. Goodman sees truth here as a system-neutral or trans-system notion by means of which descriptions can be sorted. Truth is, at least in *WWI*, not system-relative, and Goodmanian relativism amounts only to the rejection of any absolute, privileged system in terms of which all descriptions must be made.

So, for the Goodman of *WWI*, there are many ways the world is; many true descriptions of the world. But there are also many ways the world *isn't*; many false descriptions of the world. Goodman is here committed to there being system-neutral criteria by which system-relative descriptions of the world can be judged. As we are about to see, later versions of Goodmanian relativism are not so clear on this point.

WAYS OF WORLDMAKING: RIGHTNESS AND RELATIVISM

Ways of Worldmaking[8] is Goodman's most recent articulation of his long-standing view that a fundamental sort of epistemic activity is *construction* – we *make* things (objects, paintings, theories, etc.) which are both the expression and the object of judgment and understanding. We construct world-versions, or, more simply, versions, by manipulating systems in ways Goodman has long studied.[9]

Goodman argues, in *WW*, for the acceptance of alternative, conflicting true or right world-versions, but, as in earlier articulations of his relativist stance, he denies that it is his view that "everything goes":

Willingness to accept countless alternative true or right world versions does not mean that everything goes, that tall stories are as good as short ones, [or] that truths are no longer distinguished from falsehoods . . . (94)

The multiple worlds I countenance are just the actual worlds made by and answering to true or right versions. (94)

So, according to the view put forward in *WW*, versions can be true or false, right or wrong; even though truth or rightness cannot be conceived as "correspondence with a ready-made world" (94), still there are standards of rightness which it now becomes crucial to uncover and articulate. This becomes the central task of Goodman's inquiry:

> We must, then, inquire into the standards . . . of rightness of renderings of all sorts, in all media, in symbol systems of every variety. This strikes me as one of the most pressing problems of present-day philosophy . . .[10]

Goodman is thus centrally interested in distinguishing genuine from spurious worlds, true from false descriptions, right from wrong versions. The standards of rightness Goodman seeks constitute "rigorous restraints" on his "radical relativism." Our task now is to come to grips with rightness.

RIGHTNESS

The first thing to note is that rightness cannot be construed as truth. A version may be right but not true; incompatible versions may both be right; and to some right versions standards of truth are not even applicable:

> Not only does countenancing unreconciled alternatives put truth in a different light, but broadening our purview to include versions and visions that make no statements and may even not describe or depict anything requires consideration of standards other than truth. Truth is often inapplicable, is seldom sufficient, and must sometimes give way to competing criteria. (107)

The competing criteria are, of course, criteria of rightness, which outstrip considerations of truth:

> . . . with our view of worldmaking expanded far beyond theories and descriptions, beyond statements, beyond language, beyond denotation even, to include versions and visions metaphorical as well as literal, pictorial and musical as well as verbal, exemplifying and expressing as well as describing and depicting, the distinction between true and false falls far short of marking the general distinction between right and wrong versions. (109)

How then do we so distinguish? What are the "standards of acceptability" (110) which sometimes supplement, sometimes compete with, and sometimes replace considerations of truth in determining the acceptability of versions? Goodman's enigmatic solution is *fit*:[11]

> A statement is true, and a description or representation right, for a world it fits. (132)

. . . truth of statements and rightness of descriptions, representations, exemplifications, expressions . . . is primarily a matter of fit: fit to what is referred to in one way or another, or to other renderings, or to modes and manners of organization. (138)

Rightness most generally is "a matter of fit with practice." (138) Rightness of categorization is a matter of that categorization successfully organizing practice; rightness of deductive inference is a matter of fit with deductive practice; and similarly for rightness of inductive inference and sampling technique. (125–140) Likewise, the rightness of a painting, or a novel, or a dance, or a simple descriptive statement, is also a matter of fit with practice:

Whether a picture is rightly designed or a statement correctly describes is tested by examination and reexamination of the picture or statement and what it refers to in one way or another, by trying its fit in varied applications and with other patterns and statements. (139)

I must confess that I find Goodman's discussion of "fit" somewhat obscure, and not without its difficulties. What is it, I wonder, to "fit with practice"? Assuming this notion clear, doesn't this formulation give too much weight to practice? Surely there are cases in which what fits fits into a practice which is itself open to challenge. In such cases, where the fit is good but the practice is not, it seems inappropriate to label such versions "right." Consider, for example, a mathematical proof challenged on intuitionist grounds: the proof fits standard deductive practice, but that practice is itself challenged by the intuitionist. Is it not question-begging against the intuitionist to call the proof right, even though it "fits with practice"? Or consider the ill-informed auto mechanic in some remote part of the globe, whose garage is equipped with air compressor and some pneumatic hand tools, but not the pneumatic "gun" for removing/fastening lug nuts when changing tires. Her practice is of removing lug nuts by hand. She hears of, and adopts, a new wrench which improves leverage, and so makes such removal easier. Is the new method right? According to Goodman's discussion, as I understand it, it is – even though the original practice is in obvious need of the pneumatic gun, in the advent of which the new wrench would be otiose. Here again it seems unfortunate to label a version which fits with practice "right," when the practice itself is open to challenge and improvement. Indeed, it seems that anything can be right in that it can be seen to fit some practice, if we are only perverse enough, or liberal enough, in our consideration of practices. Of course Goodman allows that practice, and habit, can be amended and improved. (128, 135–7) But how do we know what constitutes improvement? In order to make such assessments, we must be able to criticize practices – for Goodman, presumably, this would be done in terms of our purposes – but then our purposes must themselves be subject to critical assessment, or else our perverse purposes would sanction our perverse practices. Moreover, to appeal to some criterion for assessing practices is to unhinge Goodman's analysis of rightness in terms of fit with practice, for that practice itself must then be right in order for fit with it to count as a condition of rightness.[12]

So I am not satisfied that Goodman's account of rightness is adequate for his purposes. At a minimum, much more needs to be said concerning rightness and fit for these notions to do the work Goodman wants them to do in his system.[13] However, whatever the inadequacies in Goodman's account of rightness, I propose to ignore them from here on in, for the most interesting questions raised by this account involve its relation to Goodman's thesis of "radical relativism under rigorous restraints." It is to a consideration of Goodmanian relativism that we now turn.

RELATIVISM

Goodman's version of relativism (henceforth "*GR*") is a unique and perplexing one. He calls it a radical relativism, which it seems to be, based as it is on his theses that facts are fabricated; that worlds are made, not found; and that contrary versions, conflicting truths, and unreconcilable theories can all be right. (109–110) Truth itself is relative, on Goodman's current picture,[14] as is rightness. (139) This is a radical relativism indeed, in which, it seems, anything goes. But Goodman places "rigorous restraints" on relativism, and emphatically denies that anything goes:

> That right versions and actual worlds are many does not obliterate the distinction between right and wrong versions, does not recognize merely possible worlds answering to wrong versions, and does not imply that all right alternatives are equally good for every or indeed for any purpose (20–21).[15]

> What I have said so far plainly points to a radical relativism; but severe restraints are imposed. Willingness to accept countless alternative true or right world-versions does not mean that everything goes, that tall stories are as good as short ones, [or] that truths are no longer distinguished from falsehoods . . . The multiple worlds I countenance are just the actual worlds made by and answering to true or right versions. Worlds possible or impossible supposedly answering to false versions have no place in my philosophy. (94)

Here *GR* seems not radical at all, since "what goes" must meet Goodman's criteria of rightness. The restraints on acceptable versions are indeed rigorous and severe.

Or are they? *GR* is restrained by criteria of rightness, but rightness is itself relative and a version can be right in one system but wrong in another. Rightness is relative to system. Systems themselves are neither true nor false, right nor wrong; they are judged solely in terms of their "efficacy in worldmaking" (129), and efficacy is itself relative to system. So rightness is relative to system, and systems cannot be wrong. Haven't the severe and rigorous restraints on relativism then been broken? If any version can be right, given an appropriate system, and if there are no constraints on appropriate systems save our ingenuity in concocting system-relative "efficacious" ones, then doesn't "everything go"? Do Goodman's restraints dissolve, so that *GR* is an extreme and radical relativism after all?

I will argue that there is a deep tension in *GR*, and that *GR* admits of both a weak and a strong (radical) version. I will argue further that the radical version of *GR* is untenable, for it succumbs to the general difficulties, articulated in earlier chapters, which defeat any radically relativist view. The task at hand is to clarify *GR* – to delineate the weak and strong versions of *GR* – and to assess those versions. For we want to know the extent to which *GR* is subject to the incoherence arguments against epistemological relativism developed in previous chapters. Is *GR* captured by *ER*, the characterization of epistemological relativism developed in Chapter 1, and so open to the incoherence arguments delineated there? Is *GR* self-defeating? Or, for Goodman, is it possible to "commense", i.e., to neutrally compare and evaluate alternative versions?

GR

Our answers to these questions will turn on the status of criteria of rightness in Goodman's view. Radical relativism of the sort captured by *ER* ultimately fails because of its denial of neutral criteria by which alternative claims or versions can be judged. Does Goodman similarly deny the existence of such criteria? Unfortunately, there is a systematic ambiguity in *WW* which makes a straightforward answer to this question impossible. *WW* actually gives us *two* relativisms: a weak one (henceforth GR_1) according to which there are version-neutral criteria of rightness by which alternative versions can be judged; and a strong one

(GR_2) according to which criteria of rightness are version-bound. Goodman thinks rightness is relative because alternative contrary versions can be right in different systems:

> . . . rightness of design and truth of statement are alike relative to system: a design that is wrong in Raphael's world may be right in Seurat's, much as a description of the stewardess' motion that is wrong from the control tower may be right from the passenger's seat . . . (139)

As he stresses throughout *WW*, though, this does not mean that "everything goes": given a system, a version's rightness is an objective matter (139), determinable by appeal to criteria of rightness.[16] Thus versions can be wrong as well as right. Criteria of rightness are taken here to be version- and system-neutral, and so capable of guiding the objective evaluation of versions. This I want to call GR_1, or weak relativism, for in no way does it preclude objective evaluation. On the contrary, it explicitly aims for such evaluation, and Goodman (as we have seen) acknowledges the establishment of such criteria as a central focus of his work. In fact, GR_1 seems to be a straightforward and natural extension of Goodman's pluralism (discussed earlier). According to GR_1, there are many ways the world is; many true descriptions of the world. But there are also many ways that the world *isn't*; many false descriptions of the world. Moreover, thanks to recourse to criteria of rightness, we can differentiate between right and wrong, and true and false, versions. Goodman here wants to say: "Since we are relativists about the appropriateness of alternative frameworks, and the truth of alternative versions, it is important to worry about and try to establish general criteria for determining the rightness of versions." GR_1, then, in its acknowledgment of version-neutral criteria by which versions can be judged, is a relativism of a much weaker sort then the relativism stemming, e.g., from Kuhn's incommensurability thesis.[17] In the present context it is crucial to note that GR_1 is *not* captured by *ER*, since, unlike the latter, the former *does* acknowledge version-neutral criteria by which alternative versions can be non-question-beggingly evaluated.

If this were all Goodman said about relativism, there would be no problem with *GR*. But there is a problem yet to be dealt with. As we have seen, Goodman escapes a radical relativism with *GR*, because according to GR_1 there are version- and system-neutral criteria of rightness by which versions can be objectively evaluated. However, can Goodman consistently hold that his criteria of rightness are in the relevant sense neutral? It appears that he cannot.

The crucial point is that the argument formulated in *WW* is itself a version concerning versions, systems, facts, etc. *WW* presents, we might say, a meta-version: a version of versions. In Goodman's meta-version, there are version-neutral criteria of rightness by which versions can be judged. But, as Goodman recognizes, his meta-version is itself only one of countless possible meta-versions. (107) So the restraints on radical relativism which keep it from being the case that "everything goes" in GR_1 are themselves relative to Goodman's meta-version. Relativity of versions re-arises at the level of meta-version. In short, it is the case that not "everything goes" *only in Goodman's meta-version*. And since Goodman allows that his is only one among incompatible meta-versions, he must also allow the equal legitimacy of meta-versions with alternative criteria of rightness, as well as meta-versions with no criteria of rightness at all. Thus Goodman's criteria of rightness cannot stand, as in GR_1, as version-neutral criteria. Rather, those criteria are relative to Goodman's meta-version, and alternative, incompatible meta-versions, which do not recognize Goodman's criteria, must be countenanced.

But of course if those alternatives are countenanced, then the rightness of versions is relative to the criteria of rightness (if there be such) employed by the meta-version from which the judgment of rightness is made. Alternative meta-versions will license alternative judgments of rightness. Some may even deny the legitimacy or meaningfulness of such judgments. And on Goodman's view all such meta-versions must be countenanced.

One might think – wrongly, as I shall now argue – that the line of argument just offered fallaciously reasons from "some" to "all."[18] For I seem to have been arguing that since Goodman recognizes some alternative versions and meta-versions, he therefore recognizes all. This, however, is not my thesis. It is rather that the restraints on radical relativism, the criteria of rightness, are incapable of ruling out any alternatives. This follows from the fact that, for Goodman, those criteria constitute bone fide criteria only in Goodman's meta-version.

Goodman claims that versions can be right or wrong, depending on their being picked out as right by the criteria of rightness of his meta-version. So consider two versions, V_1 and V_2. V_1 is picked out as right by Goodman's meta-version; V_2 is not. Goodman claims that V_1 is right, V_2 wrong. However, these judgments of rightness and wrongness must themselves be relativized to Goodman's meta-version, for Goodman recognizes that his meta-version is itself one among countless incompatible meta-versions. He cannot say that V_1 is right and V_2 wrong *simpliciter*;

he must claim, rather, that V_1 is right and V_2 wrong *relative to his meta-version*. Thus we have, not that V_1 is right, but that V_1 is right$_G$. But another meta-version (call it F) may judge V_1 and V_2 differently from Goodman's. It may, that is, judge V_1 wrong and V_2 right. The question comes: is V_1 right? The answer, from Goodman's own point of view, must be that V_1 is right$_G$ but wrong$_F$. Rightness and wrongness of versions (and meta-versions, and meta-meta-versions, etc.) are relative not simply in the pluralistic sense that multiple irreconcilable versions can all be right. Rather, rightness and wrongness of versions are radically relative in that all judgments of rightness and wrongness can be made only relative to some meta-version, and rival incompatible meta-versions will judge differently. Judgments of meta-versions fare similarly. Thus no versions can be ruled out or judged wrong; they can only be judged wrong relative to some meta-version, and may be judged right relative to some other (equally legitimate) meta-version. Thus, I am not arguing from "some" to "all." I am arguing, rather, that Goodman's account commits him to the (relative) rightness of all versions, and the ("absolute") wrongness of none.

I am not, then, arguing that since Goodman allows that multiple incompatible versions can be right (pluralism), he thereby and for that reason allows that all versions must be right. This would be fallaciously to argue from "some" to "all." I am arguing, rather, that while Goodman wants his criteria to discriminate between right and wrong versions, they are powerless to do so. They are powerless to do so because their own status renders them impotent: they are, on Goodman's own scheme, no more authoritative than rival, incompatible criteria (stemming from rival meta-versions) which offer different judgments regarding the rightness of (object-level) versions. So a version that is judged right according to Goodman's meta-version may be judged wrong from a rival meta-version. And Goodman offers no reason to take the judgments of rightness stemming from his meta-version any more seriously than the judgments of alternative meta-versions. Indeed, (in some passages) he denies that there could be such reason.

Thus my argument is not: countless alternatives, therefore anything goes. I agree that "some" does not imply "all." My claim is rather: impotent criteria, therefore anything goes. Goodman's criteria of rightness do not yield judgments of rightness and wrongness which are in any way more authoritative than opposite judgments stemming from alternative meta-versions. Hence they do not rule out any judgments. Even though Goodman's meta-version declares some versions to be wrong, they are wrong *only* from the perspective of Goodman's meta-version.

Thus on Goodman's meta-version a version cannot simply be right "for a world it fits" (132), for whether or not it fits, and indeed whether or not fitness is the proper determiner of rightness, depends on the meta-version one judges from. The rigorous restraints on radical relativism have dissolved, and radical relativism remains. Goodman is committed, then, to a radical relativism according to which neutral evaluation of versions is impossible. This strong or radical form of *GR* – call it GR_2 – is very much like the *ER*-type relativism discussed earlier. For if criteria of rightness are relative, then, while in Goodman's version it is not the case that everything goes, it nevertheless remains that in rival versions anything might go, and in any case there will be no way to judge meta-versions, and so no way to choose from among incompatible judgments of rightness which flow from incompatible meta-versions. Thus the fundamentalist, the Goodmanian, and anyone else with a meta-version will be on an equal footing with respect to judgments of rightness.

Goodman presumably wants to say that his meta-version is *right* – not just in the sense that it satisfies his criteria of rightness, but that it is in some sense cognitively superior and preferable to rival meta-versions. Indeed, this is presumably what he is arguing in *WW*. The problem is that he cannot say this, given his meta-version's insistence (so to speak) that countless alternative and conflicting meta-versions must be recognized as legitimate and cannot be appropriately evaluated by appeal to the criteria of evaluation of his own meta-version.[19]

Of course Goodman does not want this to be the upshot of his meta-version. He thinks his meta-version is better than (more right than?) the fundamentalist's (which holds "that facts are found not made, that facts constitute the one and only real world, and that knowledge consists of believing the facts."(91)) Nevertheless, this radical relativism is what Goodman's position is driven to, once it is recognized that his meta-version is itself a version, and that his account of criteria of rightness is only one among countless incompatible alternatives. Goodman offers his restraints on radical relativism, his criteria of rightness, to be version-neutral, and to pick out his version as right. But, by his own scheme, those restraints, those criteria cannot be seen as version-neutral, but rather must be seen as part of his meta-version – and so cannot non-question-beggingly pick out his version as right. Radical relativism, Goodman style (GR_2) remains.

Besides the problems just noted, there are additional difficulties with GR_2. Most obviously and most importantly, it succumbs to the incoherence arguments developed in previous chapters. For if all meta-versions are on an equal footing according to GR_2, then meta-versions which are incompatible with GR_2, and which hold that all meta-versions are not on an equal footing, are to be taken as seriously as GR_2. Thus if GR_2 is right with respect to the status of alternative meta-versions, it is wrong from the perspective of (at least some of) those alternatives. Moreover, as argued earlier, the notion of rightness is given up by an *ER*-type relativism such as GR_2, so that its own rightness or epistemic worthiness is not coherently assertible. It is worth pointing out as well that it is not clear that GR_2 can even be coherently articulated. Goodman, in order to defend GR_2, seems to have to say that it is version-neutrally the case that rightness is relative to versions. But this is incoherent, since if rightness is relative to versions, it cannot be the case that this particular claim is itself version-neutrally right. Indeed, it is not even clear that we can refer to versions or systems as we have been doing, since given GR_2 what counts as a version or system must depend on some meta-version.[20] Thus the simple articulability of GR_2 (let alone the assertability of its rightness) is problematic.

If GR_2 is impossible to articulate, let alone defend, what shall we say of Goodman's efforts in *WW* to provide a radical relativist position? The answer is clear: Goodman can escape the difficulties of GR_2 by retreating to GR_1. Since GR_1 is a position he wants to hold, all Goodman need do is patch up his system to avoid its yielding GR_2. He needs to avoid the dissolution of the "rigorous restraints" he has placed upon his relativism. That dissolution and its avoidance depend entirely on the status of those restraints: namely, the status of Goodman's criteria of rightness. If those criteria are taken to be version-bound, then the unacceptable GR_2 results, and the rigorous restraints on relativism collapse. But if those criteria are taken to be version-neutral, and to afford the possibility of objective evaluation of versions, then GR_1 results. This weaker sort of relativism – or, better, pluralism – avoids the difficulties noted thus far with GR_2. Goodman needs to take more seriously his own view that criteria of rightness afford the possibility of objective evaluation of versions.

This last point becomes obvious if we consider for a moment the nature of criteria. What is it to be a criterion? A necessary condition of something's successfully functioning as a criterion is that it discriminate or differentiate between items which meet it and items which do not. A dictionary can serve as a criterion of correct spelling at a spelling bee only because it can discriminate between words spelled correctly and words spelled incorrectly. Similarly, the U.S.D.A.'s criteria for meat-grading function as criteria because they can consistently and routinely differentiate between cuts which are "prime," "choice," "standard," "commercial," "utility," "cutter," and "canner" (in terms of fat content, degree of marbleization, tenderness, juiciness, flavor, etc.). So, a putative criterion purports to discriminate or differentiate, a successfully functioning criterion does so. Consider now Goodman's criteria of rightness. In GR_2, they fail to discriminate between right versions and wrong versions, since at best they do so only from the perspective of Goodman's meta-version, and according to GR_2 alternative contrary meta-versions, which may yield alternative judgments of rightness (or deny the legitimacy of the category altogether), are to be equally countenanced. This is analogous to our judging a spelling bee by appealing to a dictionary which countenances all spellings of every word, or to our grading a piece of meat by appealing to a grading system that allows any piece of meat to be of any grade. In all these cases, the purported criteria fail to discriminate, and so fail to function successfully as criteria. We could say, then, that in GR_2 Goodman's criteria of rightness fail to discriminate between right and wrong versions, and so fail to constitute rigorous restraints upon relativism. The criteria of rightness, in short, do not meet the criteria for criteria: they do not discriminate or differentiate, and so do not function successfully as criteria. Another way to put the point is that criteria must be taken to be version-neutral in order to have any force *as criteria* at all.[21]

Goodman clearly wants his criteria of rightness to afford a means of legitimate evaluation of versions. He does not want to be committed to the view that "everything goes," that all versions are equally acceptable. This undesired (and undesirable) result follows if we understand *GR* as GR_2, but it is avoided if we take *GR* as GR_1. Goodman has opined that he views inquiry "into the standards . . . of rightness of renderings of all sorts . . . as one of the most pressing problems of present day philosophy . . ."[22] If so, he had better be advocating GR_1. For if he is advocating GR_2, general standards of rightness are, as I have been arguing, in principle unachievable. Holding to GR_2 undermines the inquiry into standards of rightness that Goodman regards as centrally important.

Putnam, in his commentary on *WW*[23], recognizes the problem for Goodman here. Putnam worries about the case in which "right" means "right for me," and notes the possibility of a culture's holding a radically relativist position (as Putnam says, "the young in California *do* say 'true for me'").[24] And in his closing he urges Goodmanians to avoid the radical relativist position, and to work towards the establishment of "rigorous standards for distinguishing right from wrong."[25] But Goodman, in his reply to Putnam, resists Putnam's attempt to short-circuit GR_2. He writes:

> Standards of credibility do not vary with individual opinion; they are constant in the world of worlds sketched in my book but may vary from one world of worlds to another.[26]

Here Goodman wants the impossible. For if standards may vary from one world of worlds – i.e., meta-version – to another, then they may well vary with individual opinion. At least, they will if individuals espouse alternative meta-versions (worlds of worlds). In short, if meta-versions can vary with individual opinion, so, contrary to Goodman, can standards of credibility and of rightness. So Goodman's reply, offered as an attempt to discount Putnam's suggestion that *WW* offers in the end a radically relativist view in which standards vary with individual opinion, does not alleviate Putnam's worry.[27]

CONCLUSION

It must be emphasized that (at least in some passages) Goodman does take his constraints on relativism seriously, and so rejects the sort of (Kuhnian) radical relativism in which "anything goes." In contrast with Kuhn, Goodman does not regard alternative versions as incomparable.[28] This Goodman – the Goodman of GR_1 – I regard as an ally in the battle against radical relativism. My aim in this chapter, however, is to point out that other passages of *WW* seem to commit Goodman to the sort of radical relativism (i.e. GR_2) that he elsewhere rejects, and to argue that this latter sort of relativism is subject to the same sorts of difficulties as other forms of radical relativism. There is, as I have said, a tension in *WW* between relativisms of different strains and strengths. It is this ambiguity in *GR* which I have endeavored to point out; I have argued, in addition, that while one strand of *GR* is perfectly defensible, the other is not. It is this latter strand that has been the focus of my critical comments. The object of my criticism is not Goodmanian relativism as such, but only the radical form of it (GR_2) which aligns *GR* with other untenable versions of radical relativism. I am completely in sympathy with what perhaps is best called "Goodmanian pluralism." Friends of *GR*[29] generally defend the pluralist GR_1, not the radically relativist GR_2. To this extent, there is no conflict between me and those defenders of *GR*. If Goodman is committed only to GR_1 and not GR_2, there is (from my point of view) no problem with *GR*. It is only the radically relativist (and, if the arguments of earlier chapters are correct, incoherent because self-refuting) GR_2 that I am criticizing here. My view is: pluralism, yes; radical relativism, no.

If what has been said thus far is correct, then Goodman had best be advocating GR_1, not GR_2. GR_1, which is a pluralist rather than a radical relativist position, is sufficient to underwrite Goodman's central investigations into version-making through symbols and the general nature of symbol systems. In fact, Goodman had better hold GR_1, and not GR_2, if he wishes to investigate the nature of general standards of rightness of renderings, since, as we have seen, on GR_2 such standards are impossible.

WW is without doubt a rich and suggestive book, and is certain to spur research along the lines of Goodman's philosophical project. Nevertheless, as it stands there are specific difficulties with Goodman's rendition of irrealism, rightness, and relativism which call for rigorous restraint in affirming the rightness of Goodman's rendering.

CHAPTER 8

RELATIVISM AND RATIONALITY: TOWARDS AN "ABSOLUTIST" EPISTEMOLOGY

Suppose I am right that, for some or all of the reasons adumbrated thus far, epistemological relativism is untenable: suppose, that is, that the relativist position cannot be asserted or defended without being embroiled in self-contradiction. What follows? What does the rejection of relativism imply for epistemology? What must an "absolutist," i.e. non-relativist, epistemology be like?

Several of the relativists we have considered thus far have suggested that relativism is to be preferred to absolutism because the latter carries with it several ramifications for epistemology that we should be loathe to accept: for example, that absolutism entails certainty, or some necessary privileged framework, or the unrevisability of some class of statements, or dogmatism. The first task of the present chapter is to sketch the broad outlines of an absolutist epistemology and to show that none of these untoward and undesirable features are features of the absolutism sketched.

1. ABSOLUTISM AND ITS PURPORTED UNDESIRABLE RAMIFICATIONS

Edward Beach helpfully distinguishes between two conceptions of absolutism:

> *Absolutism*: The thesis that there exists a set of necessary principles which are available to knowledge and are capable of grounding an eternally valid, incorrigible belief system. These incorrigible truths can thus serve as the criteria by which to evaluate all subsidiary beliefs, including empirical, hypothetical, and perspectivally determined ones. What counts as necessarily true must be universal and absolute, for it is only be [sic] means of the absolute that even the relatively [sic] of particular perspectives can be known as such.
>
> *Objectivism*: The thesis that there exists a systematic method of reasoning and a coordinate set of beliefs embodying its principles which, despite the vicissitudes of social psychological conditioning, are accessible to knowledge and are capable of sustaining a dynamic, self-correcting belief system. These principles may contain errors or half-truths, and they may never attain to a fixed and final form. Yet insofar as (a) their consistency is publicly verifiable, (b) their development is rational, and (c) their truth-content is demonstrably greater than that of rival contenders, they do constitute reliable criteria by which to evaluate subsidiary beliefs and hypotheses. What counts as effectively true must include at least implicit judgments about the universal and objective, for it is only be [sic]

reference to an objective system of reality that particular assertions function at all as coherent cognitive acts.[1]

Beach notes that one of his purposes in introducing these sketches of alternative non-relativistic positions is "to point out that the denial of relativism does not necessarily entail [what Beach calls] absolutism."[2]

The articulations of these two non-relativistic positions are not entirely unproblematic. There is also a difficulty with the names of the two positions, at least in the present context, since the position Beach labels "Objectivism" is a species of "Absolutism" as I am using the term (and so not a contrary of it, as Beach uses the term). Nevertheless, Beach's formulations are exceedingly useful, as they serve to distinguish between tenable and untenable alternatives to relativism. In what follows, therefore, I shall consider the features of the two positions.

The first [Beach's "Absolutism"] contrasts with relativism in that it maintains that some knowledge is certain or incorrigible. In addition, this position maintains the existence of a unique, necessary privileged framework ("an eternally valid, incorrigible belief system . . .") from which all other putative knowledge-claims can be judged. Some relativists have argued that the rejection of relativism necessitates the acceptance of these features of absolutism. Thus, as we saw in Chapter 1, Field argues that the absolutist is committed to the existence of a unique privileged framework, namely, "*the true evidential system*," which is the only system which "correctly describes genuine justification"[3]; similarly, Brown claims that absolutism assumes or presupposes an infallibilist, foundationalist epistemology.[4] Thus these two relativists, at least, think that this first absolutist position follows from the rejection of relativism.

However, as Beach points out, a second alternative to relativism [Beach's "Objectivism"] is possible. This second position does not require that there be a unique privileged framework from which to assess knowledge-claims; nor does it require that knowledge be incorrigible or certain. According to this second position, fallibility and revisability reign, with respect both to putative knowledge-claims and to the evaluative criteria by which such claims are assessed. Knowledge-claims can be objectively assessed in accordance with presently accepted criteria (e.g. of evidential warrant, explanatory power, perceptual reliability, etc.), which can in turn be critically assessed. Thus an absolutist belief system can be both self-correcting and corrigible. Furthemore, judgments of knowledge-claims, as Beach suggests, requires at least implicit commitment to "absolutist" presuppositions – not regarding certainty or privilege, but rather the possibility of objective, non-question-begging judgments. We have seen this commitment several times thus far. In Chapter 1, I argued that even the relativist must make it, in order to defend relativism (thus illustrating the self-defeat of relativism, which is required to presuppose what it explicitly rejects).[5] In Chapter 7, I argued that criteria, in order to function as criteria, must likewise presuppose an objective or absolute (but not incorrigible) status. But I will not repeat those arguments here. Rather, I wish to underscore the point that the absolutist commitment to objective, non-question-begging criteria is not tantamount to epistemological assumptions of incorrigibility, certainty, or a unique privileged framework. Rather, as suggested in Chapter 1, absolutism may be fallibilist; as argued in Chapter 5 and elsewhere, absolutism requires not certainty, or "ultimate" truth, but simply that knowledge-claims be evaluated non-arbitrarily, and that they be more or less warranted on the basis of reasons which serve to (fallibly) justify their acceptance. Finally, absolutism requires not the identification of a unique privileged framework or evidential system, but (again as argued in Chapter 1) simply the possibility of objective, non-question-begging evaluation of putative knowledge-claims, in terms of criteria which are taken as absolute but which nonetheless admit of criticism and improvement. The rejection of relativism, then, entails neither certainty nor a unique privileged framework or evidential system. As Beach rightly points out, his first position is not the only version of absolutism extant. A fallibilist, non-dogmatic, self-correcting absolutism, which provides for objectivity in the evaluation of knowledge-claims and the possibility of criticism and improvement of criteria which guide such evaluation, is entirely compatible with the rejection of relativism.

The first position [Beach's "Absolutism"], which might be called "vulgar absolutism," is not necessitated by the rejection of relativism. A more defensible absolutism includes none of the features – incorrigibility, certainty, unique privileged framework, etc. – of vulgar absolutism. It is an absolutism of the non-vulgar sort that I am defending here. For, as several philosophers have pointed out, none of the features of vulgar absolutism are required by the rejection of relativism.[6]

The relativist thus gleans undeserved support from the failure to distinguish vulgar from a more defensible absolutism, along with a concomitant rejection of (vulgar) absolutism and an inference from the untenability of (vulgar) absolutism to relativism. Once the distinction between vulgar and non-vulgar absolutism is drawn, however, the inference fails, for the rejection of vulgar absolutism leaves us not with relativism, but with a choice between (at least) non-vulgar absolutism and relativism. Given the arguments adduced thus far against the latter, our choice here seems rather obvious.

Relativism also illicitly gains from the failute to pay sufficient attention to another distinction, namely that between relativism and pluralism. The pluralist does not hold, with the relativist, that there is no possibility of non-neutral or non-question-begging evaluation of alternative claims, theories, schemes, versions, or the like. The pluralist subscribes, rather, to a willingness to tolerate and utilize a diversity of ideas and approaches, while at the same time acknowledging criteria which afford the possibility of objective comparison and evaluation of the diverse alternatives tolerated and utilized.[7] The tolerance of alternatives exhibited by pluralism is admirable, for (among other things) it is quintessentially non-dogmatic, and dogmatism is yet a further undesirable feature of vulgar absolutism.

In failing to distinguish relativism from pluralism, and vulgar from non-vulgar absolutism, one may illicitly argue in relativism's defense that it has the advantage of avoiding dogmatism. I agree that such avoidance is desirable. But it is crucial to see that that advantage is a feature, not (only) of relativism, but (also) of pluralism. Moreover, dogmatism is a feature of vulgar absolutism, but not non-vulgar absolutism. So we may all agree that dogmatism is undesirable and is to be avoided. What follows is not that we should therefore reject absolutism and embrace relativism. What follows, rather, is simply that we should reject vulgar absolutism; we remain free to embrace pluralism. Furthermore, in embracing pluralism, we do not reject non-vulgar absolutism, for these two views are perfectly compatible.

One may reasonably ask, given the weaknesses of relativism, why anyone would be moved to hold it. While I cannot speak for the motivations of any particular relativist, it seems plausible that one of the main motivations[8] of many relativists is the (entirely justifiable) desire that one's epistemology avoid dogmatism, certainty, incorrigibility, a unique privileged framework, and the other evils of vulgar absolutism discussed thus far. The view that these features of vulgar absolutism are undesirable and to be avoided is laudable. It constitutes a consideration in favor of relativism, however, only if one thinks that relativism is the only alternative to vulgar absolutism. In this case, one reasons by disjunctive syllogism: either vulgar absolutism or relativism, but not the former (because of the undesirable features), hence the latter. The problem, of course, is with the first premise. Vulgar absolutism and relativism are not the only options. Once non-vulgar absolutism is distinguished from the vulgar sort, and pluralism distinguished from relativism, the desirability of avoiding the undesirable features of vulgar absolutism ceases to constitute a good argument, or a powerful motivation, for relativism. It constitutes, rather, an argument for the disjunction: relativism or non-vulgar absolutism or pluralism. And, given the arguments against relativism advanced throughout the earlier chapters, it is clear that we should eliminate from contention the first disjunct. This leaves us with either non-vulgar absolutism or pluralism. But we need not choose between these two, for they are wholly compatible. Indeed, as I argued in Chapter 7, pluralism is eminently worthy of our embrace, and Goodman[9] is quite right in urging us to accept it. Not only can we embrace pluralism and non-vulgar absolutism, we should embrace them both. But pluralism is manifestly not tantamount to relativism, and the desirability of the former in no way extends to the latter. Distinguishing between pluralism and relativism, and between vulgar and non-vulgar absolutism, thus undercuts a central motivation for relativism.

Absolutism[10] thus avoids the features which many rightly take to be undesirable. The absolutist need not, indeed should not, embrace the sort of absolutism which entails that knowledge is certain, incorrigible, or the product of a unique privileged framework. Nor need the absolutist portray the knower as dogmatic. A non-dogmatic, non-certain, corrigible, fallible, non-unique absolutism is both coherent and tenable. It is this sort of absolutism I am recommending here. The purported undesirable ramifications of absolutism are indeed undesirable. But they are not genuine ramifications of absolutism.

My point in all this is not to criticize vulgar absolutism. Contemporary epistemologists – absolutists and relativists alike – reject certainty, dogmatism, and all the other features of vulgar absolutism. My point, rather, is that the rejection of relativism does not force vulgar absolutism (as Brown, e.g., suggests) upon us. That rejection is compatible with an absolutism of a non-vulgar sort. So we can reject relativism, on the basis of the arguments presented in earlier chapters, without worrying that that rejection commits us to an equally unpalatable epistemological doctrine, i.e., vulgar absolutism.

2. ABSOLUTISM AND ITS RELATION TO OTHER CENTRAL QUESTIONS OF EPISTEMOLOGY

One might think that opting for an absolutist epistemology commits one to definite positions on a variety of other epistemological controversies. In this section, I want briefly to argue for two claims. First, it is not the case that absolutism entails an entire epistemological program. Rather, absolutism leaves open a whole range of central epistemological issues. Second, the regarding of these issues as genuine issues requires the acceptance of absolutism and the rejection of relativism. That is, absolutism is a necessary precondition of epistemological inquiry.

Does the embracing of absolutism commit one to any particular answers to standard epistemological questions? While I will not argue the point in any detail, I believe that no such commitment ensues. The absolutist can remain agnostic, or take either side, in the debate concerning (for example) theories of truth – that is, one can be an absolutist and embrace either a correspondence or a coherence theory. Similarly for theories of justification. As regards the analysis of knowledge, the absolutist *qua* absolutist can endorse the standard analysis, a causal theory, a defeasibility theory, a reliabilist theory, an explanatory theory, or virtually any other sort of theory. Likewise for the theory of evidence. Finally, and perhaps most contentiously, absolutism is compatible with both foundationalist and non-foundationalist epistemologies.[11] Absolutism requires that it be possible that claims be evaluated in a non-question-begging way, and that objective comparison of rival claims be possible. It does not require that knowledge be certain, or indubitable, or rest on considerations to which we have privileged access. For the criteria which afford objective, non-question-begging evaluation may themselves be criticized, compared with alternatives, and improved. There is no particular difficulty in holding a non-foundationalist absolutism.[12]

So the absolutist is not, in embracing absolutism, committed to particular views concerning particular central epistemological controversies. She is free to take sides on these issues as reasoned argument suggests. She is free, that is, to pursue epistemological inquiry. Moreover – and this is the second point mentioned above – only the absolutist is free to regard epistemological inquiry in this light. The relativist must regard epistemological debate as pointless, insofar as there is, for the relativist, no possibility of genuinely answering central epistemological questions. For the relativist, having given up the absolutist conception of rightness, cannot assert that foundationalism (non-foundationalism), correspondence (coherence) theories of truth or justification, causal (reliabilist, defeasibility, etc.) theories of knowledge, or the like are non-relatively right. But genuine epistemological debate does have as its aim the determination of the non-relatively right answers to these questions. The relativist cannot regard epistemological questions as serious ones which admit of non-relative examination. Thus, not only can the absolutist regard epistemological inquiry as legitimate and open, *only* the absolutist can. The relativist, *qua* relativist, cannot.[13] In the sense that general epistemological inquiry requires or presupposes absolutism, further, it appears that the relativism/absolutism controversy is prior to, or more basic than, other epistemological concerns. It is epistemological bedrock.

3. PRELIMINARY CONCLUSION

What I have offered thus far is only the briefest sketch of an absolutist epistemology. (I shall say more about what a full characterization of absolutism requires in a moment.) Nevertheless, I hope to have shown in this chapter that absolutism can take a palatable and plausible form, that it need not be saddled with the undesirable features of vulgar absolutism, and that the relativist derives no comfort from those features (since absolutism does not involve them). I hope also to have made clear that absolutism, but not relativism, allows for the rational pursuit of epistemology. I have suggested, finally, that the absolutism/relativism dispute is neither idle nor unimportant from the point of view of mainstream epistemology. On the contrary, since absolutism is necessary for, or presupposed by, general epistemological inquiry, it is essential that epistemology confront the absolutism/relativism controversy. In this sense the controversy is a – perhaps *the* – fundamental epistemological problem. It is absolutely basic to the epistemological enterprise.

Moreover, if these last points are correct, then not only is the controversy basic, but it is a requirement of epistemological inquiry that absolutism, and not relativism, be the rational outcome of the controversy. Consequently, it is a happy result – at least for those of us who wish the continued health and pursuit of epistemology – that the preceding chapters furnish compelling reasons for rejecting relativism and embracing absolutism.

4. RELATIVISM AND RATIONALITY

I cannot claim to have nailed relativism's coffin finally shut. For there remain many schools of relativistic thought that I have either only barely touched upon or entirely neglected. The enormous literature spawned by Wittgenstein concerning forms of life and the limits of reason; Winchian considerations, and social scientific, especially linguistic and anthropological, considerations which suggest the incommensurability of alternative conceptual schemes; the "strong programme" of the Edinburgh school in the sociology of knowledge: all of this literature demands explicit attention. While I firmly believe that the anti-relativist arguments offered above apply with equal force to this additional relativist literature, I cannot claim to have shown that here.

While this book has been concerned entirely with epistemological relativism, the recent revival of moral relativism by such writers as Brandt, Harman, Mackie, Williams, and Wong also deserves attention. While the establishment of some form of moral relativism would not harm this book's central thesis (since that thesis concerns only epistemological relativism), and so strictly speaking a full treatment of epistemological relativism need not involve consideration of moral relativism, I nevertheless suspect that there is much to be gained from sustained study of the latter. Moral relativism may or may not escape the sorts of arguments developed here against its epistemological cousin. (I suspect that whether or not it does depends on the viability of the epistemology of moral claims underlying moral relativism.) Pursuit of this and related topics is clearly in order. So there is much more to be said about relativism.

But there is a deeper problem remaining. I have argued that absolutism need not be vulgar; that absolutism requires neither certainty, nor an indubitable framework into which all knowledge-claims must fit, nor a set of incorrigible criteria by which putative knowledge-claims must be judged. I have urged, rather, that all that is required for absolutism is the possibility of objective, fair, neutral, non-question-begging evaluation of rival claims, in accordance with criteria which themselves admit of critical assessment and improvement. What would such evaluation, and such criteria, be like? It seems that, even if this conception of absolutism is right, the reader deserves a longer story than the one I have provided here.

I agree. I think that the longer story required is a story about reasons, or, more glorifiedly, about rationality. For in claiming that rival claims can be fairly compared, or that particular claims can be evaluated and assessed, I am claiming, in effect, that we can ask – and answer – questions concerning the (non-relative) epistemic worth of claims. We can, that is, determine the extent to which reasons for some claim offer it warrant. What is needed, therefore, is an account of the warranting force of reasons. We need a theory of rationality to secure absolutism.

We have seen in earlier chapters, especially those concerning the philosophy of science, that the notions of relativism and rationality are connected, since the embracing of relativism with respect to theory choice seemed to imply that such choice is irrational. I am suggesting here that the connection is even deeper than was suggested earlier, in that a full articulation of a theory of rationality is required for a full articulation and defense of absolutism. Such a theory must include an account of reaons, such that it is possible to establish that, independently of framework, scheme, or other relativizing factor, there is a relation *R* between claims *p* and *q* such that the claim "*p* is a (good) reason for *q*" is (absolutely) true. Thus a fully worked out absolutist epistemology hinges on the analysis of reasons, the warrant they afford knowledge-claims, and an account of the relation ". . . is a reason for. . . ."

A general theory of rationality of the sort I am here calling for would also involve systematic treatment of a variety of longstanding issues in philosophy. These include the rationality of science, the role of reason in ethics, and problems of decision theory and game theory, including the prisoner's dilemma, Newcomb's problem, etc. Such a theory must also come to grips with perhaps the most difficult problem concerning rationality of all, namely, its justification. It should answer the question, "Why be rational?" (This is turn requires consideration of the view that this is an illicit question, and that one cannot coherently ask for a (rational) justification of rationality;[14] and also of the view that the question is legitimate but cannot be answered in such a way as to justify the commitment to rationality, so that one can only have (irrational) faith in rationality.[15] I believe that both these views are mistaken, and that one *can* justify a rational commitment to rationality. But in any case, a theory of rationality of the sort I am calling for here must come to grips with the knotty problem of the epistemological status of rationality itself and our commitment to it.) This last question is forced by the relativist, since if it cannot be non-relatively answered, the relativist seems able to resuscitate her position by claiming that the reasons for abandoning relativism may be safely ignored. Thus a theory of rationality of the sort here described is necessary to establish absolutist epistemology and to put relativism finally to rest.

I therefore explicitly acknowledge that the characterization of absolutism offered in this brief concluding chapter is inadequate. It is far too superficial to establish the tenability of absolutism. I offer it only to suggest the direction in which work toward articulating a viable absolutism should go. In particular, I want to urge that that further articulation should involve a general treatment of rationality, and that such a general treatment, if successful, would provide all the basis for nonrelative judgments that absolutism requires. Seeing that absolutism hinges on the theory of rationality puts the absolutism/relativism controversy in a new light, and points the way both to a deeper understanding of the controversy and to its ultimate resolution. Thus, while this chapter is not much more than a promissory note, I hope it is an instructive note. At least it specifies the terms of payment to be provided if the obligation is to be met.[16]

NOTES

NOTES TO INTRODUCTION

[1] Meiland, 'Concepts of Relative Truth'; Meiland, 'Is Protagorean Relativism Self-Refuting?'; Meiland, 'On the Paradox of Cognitive Relativism'; Meiland and Krausz, eds., *Relativism*.
[2] Field, 'Realism and Relativism'.
[3] Popper, *The Open Society and Its Enemies, Volume 2*; Popper, 'Normal Science and its Dangers'; Popper, 'The Myth of the Framework'.
[4] Davidson, 'On the Very Idea of a Conceptual Scheme'.
[5] Quine, 'On Empirically Equivalent Systems of the World'.
[6] Cf. Grandy, ed., *Theories and Observation in Science*.
[7] For a general review of this literature and a defense of the distinction, cf. Siegel, 'Justification, Discovery and the Naturalizing of Epistemology'.
[8] For an excellent discussion of the fall of positivism, cf. Suppe, *The Structure of Scientific Theories*; also Brown, *Perception, Theory and Commitment*.
[9] Kuhn, *The Structure of Scientific Revolutions*.
[10] Meiland, 'Kuhn, Scheffler, and Objectivity in Science'.
[11] Doppelt, 'Kuhn's Epistemological Relativism: An Interpretation and Defense'; Doppelt 'Relativism and Recent Pragmatic Conceptions of Scientific Rationality'; Doppelt, 'Relativism and the Reticulational Model of Scientific Rationality'; Doppelt, 'The Philosophical Requirements For an Adequate Conception of Scientific Rationality'.
[12] Brown, *Perception, Theory and Commitment*.
[13] Stegmüller, *The Structure and Dynamics of Theories*.
[14] Toulmin, *Human Understanding*.
[15] Laudan, *Progress and Its Problems*.
[16] E.g., Goodman, 'The Way the World Is'; Goodman, *The Structure of Appearance*.
[17] Goodman, *Ways of Worldmaking*.

NOTES TO CHAPTER 1

[1] Plato, *Theaetetus*, 152a.
[2] Ibid., 160c.
[3] Ibid., 166d–167d.
[4] Ibid., 170a.
[5] Much of what follows in this chapter is concerned with the status of this relativizing phrase.
[6] Thus I will not be considering the entire range of Socrates' arguments against Protagoras. For a fuller discussion cf. M.F. Burnyeat, 'Protagoras and Self-Refutation in Plato's Theaetetus', and Burnyeat, 'Protagoras and Self-Refutation in Later Greek Philosophy'.

Of particular note is Burnyeat's "triple-sequence" analysis of the Socratic response to Protagoras.

[7] *Theaetetus*, 162a.

[8] Ibid., 169d–171c. Socrates raises an additional problem for Protagoras here in his pointing out an apparent inconsistency between Protagoras holding both that some are wiser than others and that no one thinks falsely. (Cf. esp. 169d, with reference to 167b–c.) But I shall forego consideration of this problem, since the Protagorean relativist could relinguish the former claim. The problem therefore does not count against the Protagorean relativist as such.

[9] Ibid., 170e–171a.

[10] Ibid., 171a. Here begins a famous difficulty for the Socratic argument. Socrates deletes the relativizing phrase 'for . . .' after 'true', thus unfairly characterizing Protagoras' position. This opens the door to the charge that Socrates has begged the question against Protagoras by assuming an 'absolute' conception of truth, according to which a claim's truth is independent of a person's belief in its truth, which the relativist rejects. The Protagorean counter-argument, which denies the intelligibility or rightness of the absolute conception and offers instead a relative conception of truth, as well as the question-begging charge, will be considered in detail in the following sections.

[11] Ibid., 171b. Cf. also 179b.

[12] Ibid., 171c.

[13] Socrates also objects to the Protagorean doctrine on the grounds that the truth values of certain propositions, e.g. those which make assertions about the future, are independent of opinions held prior to the future event or state-of-affairs. If *A* believes that it will rain in Katmandu on January 1, 2990, and *B* believes the contrary, their opinions are not both true, as Protagorean relativism holds, but one of them is false; furthermore, their truth value depends not on the beliefs but on the weather in that place at that time. In such cases, at least, man is not the measure of all things. While this argument does seem to show that the truth values of at least some propositions are independent of persons' beliefs – and so, that Protagorean relativism is false – it does not make any claims regarding the incoherence of that doctrine, and so will not be considered further here.

[14] The following discussion borrows heavily from my "Epistemological Relativism in its Latest Form"; and my "Relativism Refuted."

[15] In addition to the papers cited in note 14, cf. also Chapter 3, below, regarding Kuhnian relativism; and my 'Rationality, Talking Dogs, and Forms of Life' regarding Wittgensteinian relativism.

[16] This follows only if we interpret truth absolutely, so that the consequent of this conditional reads: "then, if *ER* is *absolutely* true, (at least according to that set of standards $s_1, \ldots s_n$) *ER* is false." This of course will not be accepted by the Protagorean relativist who denies the viability of the notion of absolute truth, and who consequently accuses the absolutist (Socrates) of begging the question. This move is considered in some detail below. I thank Ed Erwin, Bill Sewell and Catherine Elgin for their comments on this point.

[17] A very similar point is made by Edward Beach in his "The Paradox of Cognitive Relativism Revisited", esp. p. 15.

As should be clear from the text, this argument does not show that *ER* is itself self-contradictory or incoherent. It could be true that claims admit only of relative evaluation. What is incoherent is the effort to *defend ER*, i.e. the conjunction of *ER* and the thesis that *ER* is rationally justifiable. (For rational defense requires appeal to non-relative reasons.) Thus I am not claiming, with my reconstruction of Socrates' argument, that *ER* is itself incoherent or necessarily false. Rather, as will become clear below, I am claiming that relativism is *impotent* in the sense that defense of it necessarily involves self-contradiction. I am grateful to Ed Erwin for discussion of this point.

In fact some "defenders" of relativism do seek to defend relativism only relativistically, and claim that so doing does not trivialize their doctrine. I consider this claim below.

18 Harold I. Brown, "For a Modest Historicism", pp. 549–550.

19 Ibid., p. 541.

20 Ibid., p. 550.

21 It is perhaps worth noting that Brown's confusion in the article presently under consideration carries over into some of his other writings. In his recent book *Perception, Theory and Commitment: The New Philosophy of Science*, Brown encounters similar difficulties in his analyses of knowledge, fallibilism, relativism, truth, and certainty. Cf. Chapter 5, below.

22 Jack W. Meiland, 'Is Protagorean Relativism Self-Refuting?'; Meiland, 'Concepts of Relative Truth'; Meiland, 'On the Paradox of Cognitive Relativism'; and Meiland, 'Cognitive Relativism: Popper and the Argument From Language'. Cf. also Jack W. Meiland and Michael Krausz, eds., *Relativism: Cognitive and Moral*.

23 Meiland, 'On the Paradox of Cognitive Relativism', p. 121, emphasis in original. Cf. in this regard James N. Jordan, 'Protagoras and Relativism: Criticisms Bad and Good'.

24 And in any case this is a move which is contrary to the spirit of relativism, and makes that doctrine less challenging and interesting – as Meiland notes, "On the Paradox of Cognitive Relativism", p. 119.

25 'Concepts of Relative Truth', p. 571. Note that it is possible to develop a coherent concept of relative truth that is not independent of the concept of absolute truth – much as it is possible to define relative space in terms of absolute space. The point here is not that the concept of relative truth is necessarily incoherent if it relies on the concept of absolute truth; it is rather that such a concept could not be used in an effort to establish relativism, which denies the legitimacy of the concept of absolute truth. For the relativist cannot rely on the concept of absolute truth in order to deny the legitimacy of that concept. So it is not that a concept of relative truth which relies on the concept of absolute truth is automatically, because of that reliance, incoherent; it is rather that such a concept cannot do the work the relativist wants it to do. I am grateful here to discussions with Ted Lockhart and Donald Provence.

26 Meiland, 'Concepts of Relative Truth', p. 571, emphasis in original.

27 Ibid., p. 571.

28 Ibid., pp. 571–572.

29 Ibid., p. 574.

30 Actually this is not quite right. The *metaphysical* relativist must deny that there is a way the world is, independent of statements and W's; the *epistemological* relativist must deny only that one can *know* the way the world is, independent of statements and of W's. This distinction between metaphysical and epistemological relativism is an important one that deserves more attention that I can give it here. I am grateful to Bruce Suttle, Bill Sewell, and Harold I. Brown for pointing out to me its relevance in the present context. I should like to note, however, that it does not affect the issue being treated in the text. The

argument that Meiland's three-term relation collapses into a two-term relation still goes through.
[31] Cf. Nelson Goodman, *Ways of Worldmaking*. Goodman's unique brand of relativism is considered in Chapter 7, below.
[32] Meiland, "Concepts of Relative Truth", p. 571.
[33] Ibid., p. 574.
[34] Ibid.
[35] Recall that W is not just a letter here; it is a placeholder for persons, world-views, situations, etc.
[36] To paraphrase Meiland, Ibid., p. 574.
[37] Meiland recognizes this point is another context when he writes: "'Relative truth' is a form of truth; the expression 'relative truth' is not a name for something bearing little relation to our ordinary conception of truth." 'Introduction', in Meiland and Krausz, eds., *Relativism*, p. 4.
[38] As Passmore writes, taking '*p* is true for X' as 'X thinks *p* is true' "at once raises the question whether it *is* true." John Passmore, *Philosophical Reasoning*, p. 67.
[39] Meiland's discussion focuses on the charge as it is developed by Roger Trigg, *Reason and Commitment*. The discussion occurs in 'Concepts of Relative Truth', pp. 577–580. Many other writers have considered this charge as well. For further discussion cf., e.g., Chris Swoyer, 'True For', pp. 85 and 94; and Burnyeat, 'Protagoras and Self-Refutation in Plato's Theaetetus', p. 174.
[40] 'Concepts of Relative Truth', p. 578, emphasis in original. It is worth pointing out that Plato raises this objection as well. Cf. *Theaetetus*, 183b.
[41] 'Concepts of Relative Truth', p. 579, emphasis in original.
[42] 'Introduction', in Meiland and Krausz, eds., *Relativism*, p. 4. Cf. also p. 82. But see Meiland's opposite reading of the Measure Doctrine, 'Is Protagorean Relativism Self-Refuting?', p. 58.
[43] 'Concepts of Relative Truth', p. 579.
[44] Ibid., p. 580, emphasis in original.
[45] Ibid.
[46] Another recent discussion of relative truth is that of Chris Swoyer ('True For', in Meiland and Krausz, eds., *Relativism*, pp. 88–108). While space forbids detailed consideration of Swoyer's excellent and provocative paper here, some comment is necessary. Swoyer offers what he takes to be "a serviceable sketch of the notion of relative truth" (p. 98), according to which *p*'s truth is relative to the world as conceived by a framework, so that if *p* corresponds to the facts as constituted in a framework F, then *p* is true relative to F, or true for a user of F. Swoyer's sketch is useful, but it fails in the end to advance the relativist's case, for two reasons. First, in relativizing truth to frameworks, and in holding that the world is itself framework-relative, *p*'s relative truth amounts simply to *p*'s being accepted by that framework. While Swoyer argues that his account avoids trivializing the notion of relative truth by reducing it to mere belief, and so that "it saves such obvious phenomena as the fact that someone's believing something is neither a necessary nor sufficient condition for its being true . . ." (p. 98), an analogous problem arises for Swoyer's account. For while Jones' belief that *p* is not sufficient to establish the relative truth of *p* in F (since Jones might be mistaken that *p* corresponds to the facts as construed by F), so that *p*'s truth for Jones does not amount simply to Jones' belief that *p*,

nevertheless on Swoyer's account *p*'s relative truth *is* guaranteed by the appropriate judgment from F. If *p* corresponds to the world as constituted by F, then *p* is true for users of F. There is no room on Swoyer's account for F to be mistaken, nor is there any way to assess or criticize F's construal of the world. Consequently, "F holds that *p* corresponds to the facts" is analogous to "Jones believes that *p*" – F's verdict is sufficient to establish *p*'s relative truth. Thus Swoyer's account does not save "such obvious phenomena" as the fact that a framework's judging *p* to be true is also, like an individual's belief, "neither a necessary nor sufficient condition for its being true." Swoyer's account merely substitutes the idiosyncracy and arbitrariness of judgments or verdicts by a framework for that of beliefs by an individual or group. Relative truth comes simply to "what a framework holds."

Second, Swoyer argues that his analysis comes down *against* the possibility of truth being relative in a strong sense, i.e., that there could be things which are true in one framework but false in another (though he holds that his analysis "is compatible with the view that truth might be relative in a weak sense, that is, there could be things that were true in one framework which were not expressible and a fortiori not true in another" (p. 105). Thus Swoyer's analysis does not, by his own account, support a strong relativism of the sort defended by Meiland or Protagoras.

While I welcome this latter conclusion of Swoyer's, I believe that it leads to an inconsistency in Swoyer's overall account. The same reasons Swoyer appeals to in rejection of a strong relativism – which focus on the inadequacies of radical incommensurability – also demand a rejection of the framework-relativity of worlds which is an essential component of his account of relative truth. Thus Swoyer's account of relative truth not only does not support a Meilandean or Protagorean relativism; as it stands it does not support any sort of relativism, for the inconsistency just noted renders the sketch of relative truth he offers untenable. The framework-relativity of worlds and the uncriticizability of frameworks are both more problematic than Swoyer realizes; they are also incompatible with the notion of a framework-independent world which Swoyer acknowledges "must be postulated if the relativist's picture is to avoid the slide into subjective idealism" (p. 97).

47 Below I argue that a coherent concept of relative truth fails to avoid the *NSBF* argument for relativism's incoherence as well.

48 I ignore here the troublesome (for the relativist) point that even this notion of registering a belief as a belief seems to demand a rejection of relativism. Is it, after all, only relatively true that (for example) Reagan believes in the legitimacy of prayer in the public school? Is this not true even for those who do not believe that Reagan believes this?

49 Here we see why Brown is mistaken in claiming that relativism "affirms my right to reject competing sets of presuppositions as false." (Cf. text accompanying footnote 20 above.) All relativism affirms is the right to regard competing sets of pressuppositions as *relatively* false. And since the relativist cannot claim that her own presuppositions are, in being true-for-her (and so, relatively true), cognitively or epistemically superior or more worthy of belief than competing sets of presuppositions which are false-for-her but true-for-someone-else (and so also relatively true), relativism does not affirm the right Brown claims. Here again is the penalty exacted by relativism's impotence.

50 Maurice Mandelbaum, 'Subjective, Objective, and Conceptual Relativisms', p. 403, emphasis in original. Mandelbaum's thesis of the self-limitation of relativism is weaker than the thesis of self-refutation (although Meiland regards them as equivalent), but his

case is powerfully made, and his argument that relativism is self-limiting because the evidence adduced to support relativism must be regarded non-relativistically in order to be properly regarded *as evidence* offers additional reason for taking relativism to be incoherent. It is related to, or is a version of, the argument advanced above that relativism must be defended non-relativistically, and that a "relative defense" is no defense at all.

[51] This passage is taken from the editor's introduction to Mandelbaum's article, in Meiland and Krausz, eds., *Relativism*, p. 32. I am grateful to Roy Mash for calling the passage to my attention. Cf. also 'On the Paradox of Cognitive Relativism', pp. 125–126, and 'Is Protagorean Relativism Self-Refuting?', p. 68.

[52] 'Concepts of Relative Truth', p. 580.

[53] 'On the Paradox of Cognitive Relativism', p. 126, emphasis added. I take the word "show" in these passages to be tantamount to "rationally establish."

[54] 'Is Protagorean Relativism Self-Refuting?', p. 68.

[55] 'Is Protagorean Relativism Self-Refuting?', p. 54.

[56] Swoyer, 'True For', p. 95.

[57] Jordan, 'Protagoras and Relativism: Criticisms Bad and Good', pp. 10–11. Cf. also pp. 14 and 15. It should be noted that Jordan ultimately argues for relativism's incoherence, but on grounds he regards as different from Socrates'.

[58] Burnyeat, "Protagoras and Self-Refutation in Plato's Theaetetus", pp. 174–175 and throughout. Burnyeat cites several other critics of Socrates' dropping of the qualifier "for . . .", esp. Vlastos, in these pages and in his footnote 4, on p. 174. But it must be pointed out that Burnyeat offers an analysis of the Socratic response to Protagoras which does not conclude that Socrates begs the question, but rather that Protagorean relativism is indeed incoherent, and that Socrates' arguments are ultimately grounded appropriately:

> No amount of maneuvering with his relativizing qualifiers will extricate Protagoras from the commitment to truth absolute which is bound up with the very act of assertion. To assert is to assert that p – as Passmore puts it, that something is the case – and if *p*, indeed if and only if *p*, then *p* is true (period). This principle, which relativism attempts to circumvent, must be acknowledged by any speaker. How clearly Plato saw that, I hesitate to say. But at some level it is surely what he is reacting to (p. 195).

This suggestion, that the very act of assertion of relativism commits the relativist to absolute truth, is a familiar one, and is related to the argument advanced earlier that relativism cannot be defended relativistically. In addition to Burnyeat's discussion, cf. that of Passmore, referred to by Burnyeat, and Jordan, 'Protagoras and Relativism: Criticisms Bad and Good'.

[59] Cf. Burnyeat's compelling analysis, the conclusion of which is cited in footnote 58 above. But note Meiland's rebuttal, in 'Is Protagorean Relativism Self-Refuting?', pp. 56–63.

[60] It is perhaps worth noting that Meiland finds Burnyeat's interpretation of Plato 'very persuasive as an interpretation of what Plato actually meant." ('Is Protagorean Relativism Self-Refuting?', footnote 13, pp. 56–57.)

[61] Thus the *NSBF* argument, as well as the *UVNR* argument, gives rise to the difficulty the relativist faces regarding the defense of relativism.

[62] And in any case it must be recognized that the charge of question-begging has only been leveled at (Socrates' version of) the *NSBF* argument; the *UVNR* argument is entirely unscathed by this charge. The relativist could try to extend the charge to the latter argument by claiming that that argument assumes a notion of 'absolute rightness', but the charge is forceful only if the relativist, in making the charge, assumes the very same thing

(that is, that it is right (absolutely or *simpliciter*) that the absolutist makes the assumption in question).

[63] Hartry Field, 'Realism and Relativism'.

[64] Hilary Putnam, *Reason, Truth and History*. Putnam's discussion of relativism occurs at pp. 119–124; cf. also pp. 157 and 161–162. Putnam also criticizes relativism in his second Howison Lecture, "Why Reason Can't be Naturalized", pp. 3–23. Cf. pp. 7–14.

As several writers have pointed out, it is not clear that the positive view of internal realism Putnam defends is not itself relativistic, and thus inconsistent with his rejection of relativism. (Cf. in this regard Field, 'Realism and Relativism', p. 563, note 12; also Eric Matthews' review of *Reason, Truth and History*, esp. pp. 115–116). I believe that this is indeed a serious difficulty for Putnam, but space forbids detailed consideration of the point here.

[65] Field, 'Realism and Relativism', p. 562.

[66] Ibid., p. 562. Field argues that this distinguishes his version of relativism from Protagoras', which Field agrees is incoherent.

[67] Ibid., p. 563.

[68] Ibid., emphasis in original. Note that this version of epistemological relativism neatly matches the characterization, *ER*, given earlier in this chapter, at least with respect to justification.

[69] Ibid., p. 564, emphasis in original.

[70] Field of course accepts this distinction and utilizes it throughout his argument for relativism.

[71] Ibid., p. 564, emphasis added.

[72] And Field explicitly denies that they can, Ibid., p. 566.

[73] This, incidentally, is all that the absolutist need maintain – not, as Field intimates, that there must be a *true* evidential system (Ibid., p. 563). Cf. Chapter 8, below.

[74] Personal Communication.

[75] I note in passing that judgments regarding the reliability of alternative *E*'s will inevitably be relative, however factual a property reliability is. I am grateful here to conversation with Ed Erwin.

NOTES TO CHAPTER 2

[1] Friedel Weinert, 'Contra Res Sempiternas', p. 377.

[2] Karl Popper, 'The Myth of the Framework', p. 23.

[3] 'Introduction', in Krausz and Meiland, eds., *Relativism: Cognitive and Moral*, p. 8.

[4] They also make clear that epistemological relativism may be taken as a doctrine about truth *or* as a doctrine about standards of justification or assessment. Cf. Chapter 1.

[5] This formulation brings to mind Quine's well-known view, central to his argument for naturalizing epistemology, that there simply is no "first philosophy" stance outside of science from which to judge science. For critical discussion of this view and of the project of naturalized epistemology, cf. my 'Empirical Psychology, Naturalized Epistemology, and First Philosophy'.

[6] Karl Popper, 'Normal Science and its Dangers', pp. 56–7. Emphases in original. Popper repeats the position with specific attention to Whorf and the "prison of language" in 'The Myth of the Framework', p. 38.

[7] Quoted in Popper, *The Open Society and Its Enemies*, p. 387. In 'The Myth of the Framework', Popper cites the continuation of the Xenophanes fragment to suggest that Xenophanes was in fact an originator of the Popperian view of rationality as a matter of conjecture and refutation.

[8] *The Open Society and Its Enemies*, p. 387.
[9] 'The Myth of the Framework', p. 23.
[10] Ibid., emphasis added.
[11] Ibid., p. 45.
[12] Ibid., p. 46, emphases in original.
[13] As his footnote 21 suggests.
[14] 'Normal Science and Its Dangers'; *The Open Society and Its Enemies*, Addendum 1.
[15] Donald Davidson, 'On the Very Idea of a Conceptual Scheme', p. 5.
[16] Ibid., p. 7.
[17] Ibid., p. 12.
[18] Ibid., pp. 14–17.
[19] Ibid., p. 17.
[20] Ibid., p. 20. I have passed over Davidson's rejection of conceptual relativism stemming from partial failure of translatability, which is based on his well known views regarding the necessity of charity in translation. A position very close to Davidson's is taken by Barry Stroud, 'Conventionalism and the Indeterminacy of Translation', esp. p. 92. It is interesting to note that Stroud offers his view as Quinean, while Davidson offers his view (which is nearly identical with Stroud's) as non-Quinean, since he regards Quine as holding to the "third dogma" of a sharp scheme/content distinction.
[21] Nicholas Rescher offers a similar analysis of Davidson's "translation argument" in his 'Conceptual Schemes', esp. pp. 324–326.
[22] I here leave unchallenged Davidson's views on translation, analyticity, truth, charity, meaning, and the like. While I do not regard these central aspects of Davidson's philosophy as unproblematic, I pass over them here in order to focus more clearly, in the next section, on the anti-relativist conclusion Davidson draws.
[23] I note here the rather different dissent offered by Quine. Quine tries to save empiricism from Davidson's rejection of the "third dogma" of a sharp scheme/content distinction by maintaining a role for scheme-neutral experience "as a basis not for truth but for warranted belief." Cf. Quine, 'On the Very Idea of a Third Dogma'.
[24] Popper, 'Normal Science and its Dangers', p. 56. This of course begs the question against Davidson, who would deny that these constituted different schemes precisely because they are intertranslatable.
[25] Chris Swoyer, 'True For', pp. 88–9. For clarity, I have deleted Swoyer's references to Davidson, to the immediately preceding citation from Popper, and to Kuhn, Feyerabend and Winch as relativists who would deny the Davidsonian criterion.
[26] Rescher, *op. cit.*, esp. pp. 326 ff.
[27] I especially dissent from Rescher's discussion, which defends alternative conceptual schemes by appeal to a sort of Goodmanian "irrealism." Cf. Chapter 7, below.
[28] Weinert discusses this case in 'Contra Res Sempiternas', pp. 384–6. Of course many other examples, including those Davidson discusses, could be used here as well.
[29] Ibid., p. 386. I note here that Weinert defends a weak version of relativism (p. 390) which, though relatively innocuous, I refrain from endorsing.
[30] For consideration of the parallel dispute concerning the Kuhnian claim that standards are internal to paradigms, cf. Chapter 3 below.
[31] And it is clear that Davidson does intend his analysis to defeat the possibility of conceptual relativism. Cf. 'On the Very Idea of a Conceptual Scheme', p. 5.
[32] As Swoyer suggests. Cf. text accompanying note 25 above. I am grateful to Roy Mash

for conversation and for the opportunity to read some of his unpublished manuscripts concerning Davidson and relativism. He will surely recognize (perversions and distortions of) some of his ideas in the preceding paragraphs.

[33] I ignore here the interesting suggestion made by Strawson that Davidson fundamentally misconceives the scheme/content distinction. Cf. Strawson's 'Review of *Transcendental Arguments and Science*', p. 48.

[34] I regret that space precludes consideration of the convolutions gone through by Richard Rorty, who takes Davidson to have demonstrated the incoherence of relativism, yet builds upon Davidson's argument to develop a view which is thoroughly relativistic.

[35] W.V. Quine, 'On Empirically Equivalent Systems of the World', pp. 327–8.

[36] Quine's own commitment to relativism, given the above argument but also his views concerning the immanence of truth and ontological relativity, is obscure. Quine tries to articulate the ways in which his views commit him to relativism – in my view not helpfully – in his 'Relativism and Absolutism'.

[37] A discussion of framework relativism would not be complete without at least brief mention of Wittgenstein and "forms of life." For, on a certain interpretation of that notion, forms of life are tantamount to frameworks and guarantee the boundedness of judgment characteristic of framework relativism. On this interpretation Wittgenstein is a thoroughgoing framework relativist. Such a view, characteristic of a certain strand of Wittgenstein scholarship, may be found (e.g.) in Roger Trigg's *Reason and Commitment*.

I make no claim to any special insight into Wittgenstein, and offer no contribution to Wittgenstein scholarship. Whether or not Wittgenstein is properly taken to be a framework relativist I cannot say. But I do wish to make two points. First, many commentators do not read Wittgenstein this way, but rather interpret "forms of life" in such a way that Wittgenstein does not turn out to be offering a framework relativist view. Among such interpretations are Jonathan Lear, 'Leaving the World Alone', Part I; Max Black, '*Lebensform* and *Sprachspiel* in Wittgenstein's Later Work'; Peter Barker, 'Uncle Ludwig's Book About Science'; John H. Whittaker, 'Language-Games and Forms of Life Unconfused'; and Lawrence Hinman, 'Can a Form of Life Be Wrong?'.

Second, if Wittgenstein is properly taken to be a framework relativist, than the arguments developed thus far count as much against him as against any other of that ilk. A preliminary case against a Wittgensteinian framework relativism is offered in my 'Rationality, Talking Dogs, and Forms of Life'. There I argue, roughly, that either forms of life can overlap and are not mutually exclusive, in which case no serious relativism ensues; or that forms of life cannot overlap and are mutually exclusive, in which case relativism follows but is open to the *reductio* arguments which are developed in this and the previous chapter. Cf. also Trigg's analysis, *op. cit.*

NOTES TO CHAPTER 3

[1] Kuhn, *The Structure of Scientific Revolutions*, p. viii. This section of exposition of Kuhn's basic view will seem unnecessary to many, since his views are so widely known. Nevertheless, because of the controversy concerning his actual position and his commitment to relativism, discussed below, it will facilitate later discussion to have at hand a description of the general position Kuhn's book espouses and of the way that that position has been taken by some of Kuhn's critics to be relativistic. Readers familiar with the

Kuhnian view and its relativistic interpretation are encouraged to skip ahead to the following section of the chapter.
[2] Ibid., p. 10.
[3] Ibid., p. 10.
[4] Ibid., p. 19.
[5] Ibid., p. 11.
[6] Ibid., pp. 36–42.
[7] Ibid., p. 122.
[8] Ibid., p. 5.
[9] Ibid., p. 23.
[10] Ibid., pp. 23–24.
[11] Ibid., p. 22.
[12] Ibid., pp. 52–3.
[13] Ibid., pp. 5–6.
[14] Ibid., p. 91.
[15] Ibid., pp. 74–75.
[16] Ibid., p. 77.
[17] Ibid., p. 64.
[18] Ibid., p. 103.
[19] Ibid., p. 92.
[20] Philosophers of science have traditionally taken the view that scientific knowledge grows cumulatively, with each new theory preserving the knowledge already accounted for by previous theories as well as adding new knowledge to the already existing store.
[21] Ibid., p. 103.
[22] Ibid., pp. 109–110.
[23] Ibid., p. 122.
[24] Ibid., p. 150.
[25] Ibid., p. 94.
[26] Ibid., p. 94.
[27] Ibid., p. 148.
[28] Ibid., p. 148.
[29] Ibid., p. 149.
[30] Ibid., p. 148.
[31] Ibid., p. 149.
[32] Ibid., p. 150.
[33] Ibid., p. 151.
[34] Ibid., p. 158.
[35] Ibid., p. 158.
[36] Ibid., p. 153.
[37] Ibid., p. 169.
[38] Some (including Kuhn himself, as we shall see,) dispute this irrationalist interpretation of Kuhn. Gary Gutting, for example, in the introduction to his *Paradigms and Revolutions*, pp. 4–9, chastises this interpretation of Kuhn as it is developed by Israel Scheffler in his *Science and Subjectivity*. But Gutting's discussion takes no notice of the absolutely central Kuhnian notion of incommensurability, which is the rock upon which rests the irrationalist interpretation of Kuhn developed here and by Scheffler. In any case, see below for detailed analysis of the irrationalist interpretation of Kuhn.

Note, incidentally, that Kuhn's conception of the relativity of scientific knowledge coincides neatly with the general conception of epistemological relativism (ER) provided in Chapter 1.

[39] On the interpretation being offered here, "incommensurability" is best understood as an epistemological thesis concerning the impossibility of objective evaluation of scientific knowledge-claims, rather than as a thesis of philosophy of language or the theory of meaning. Of course proponents of the "incommensurability thesis" often support that thesis by appeal to considerations of meaning. But we should not confuse the nature of the thesis with the nature of the support for the thesis.

[40] Other criticisms focus on Kuhn's derivation of incommensurability from his view on *meaning-variance*, i.e., his view that scientific terms change meaning from theory to theory. I shall not discuss this aspect of Kuhn's view in this chapter, for it has been dealt with at length and, in my opinion, convincingly refuted by Dudley Shapere ('The Structure of Scientific Revolutions', 'Meaning and Scientific Change'), Israel Scheffler (*Science and Subjectivity*, chapter 3), and Carl R. Kordig (*The Justification of Scientific Change*, chapters 2 and 3). Kuhn has also presented incommensurability in terms of "radical translation" ('Postscript-1969,' pp. 200–4; 'Reflections on My Critics', pp. 267–77; 'Notes on Lakatos', p. 146; and *The Essential Tension*, pp. xxii–xxiii and 338–9). While I shall not discuss this at present, I believe that Kuhn's conceiving of scientists of competing paradigms as being engaged in radical translation is confused, and crucially depends on his confused discussions of meaning variance just mentioned. For brief comments on Kuhn's conception of radical translation and its relevance to incommensurability, see Shapere, 'The Paradigm Concept', p. 708. For more recent discussion of meaning change, theory change, and incommensurability, all of which cast serious doubt on the Kuhnian position, see Kitcher, 'Theories, Theorists, and Theoretical Change'; Moberg, 'Are There Rival, Incommensurable Theories?'; and Levin, 'On Theory-Change and Meaning-Change.' Finally, for a general analysis of these issues and of Kuhn which is very much along the lines of that offered here, cf. Newton-Smith, *The Rationality of Science*, esp. chapters 5 and 7.

[41] Scheffler, *Science and Subjectivity*, p. 84.

[42] Kordig, *The Justification of Scientific Change*, p. 106.

[43] Kuhn's earliest published replies to his critics are his 'Postscript-1969', and his 'Reflections on My Critics'. The present section of this chapter, dealing with Kuhn's initial responses to his critics, will focus solely on these two essays. Later replies to his critics will be discussed in subsequent sections.

[44] 'Reflections on My Critics', p. 234.

[45] 'Postscript-1969', pp. 198–9, emphasis Kuhn's. Kuhn's later replies echo these earlier ones. See, for example, *The Essential Tension*, pp. 336–9, esp. p. 337.

[46] 'Postscript-1969', p. 199. For further discussion of this point vis-a-vis Kuhn and his critics, see Scheffler, 'Vision and Revolution: A Postscript on Kuhn', p. 369.

[47] 'Postscript-1969', pp. 199–200, emphasis added.

[48] Ibid., p. 200.

[49] Ibid., p. 262.

[50] Ibid., p. 185.

[51] Shapere, 'The Paradigm Concept,' p. 708.

[52] Ibid., pp. 707–8.

[53] 'Reflections on My Critics', p. 234; 'Postscript-1969', pp. 198–9.

[54] 'Postscript-1969', p. 200.
[55] Ibid., p. 204.
[56] Shapere, 'The Paradigm Concept', p. 709.
[57] This is the conclusion that Scheffler draws. Cf. *Science and Subjectivity*, chapters 2–4, esp. p. 82. Note also Scheffler's reference to Shapere here, and Kordig, *The Justification of Scientific Change*, p. 100. Laudan argues similarly ('Two Dogmas of Methodology', p. 596) that "incommensurability of theories at the object-level does not entail incomparability at the meta-level." See also Watkins, 'Against "Normal Science"', p. 36.
[58] In am indebted to Carl R. Kordig for extremely helpful discussion concerning the relationship between incommensurability and incomparability. See his *The Justification of Scientific Change*, pp. 52–57, for criticism of Feyerabend's effort to maintain both incommensurability and comparability.
[59] 'Theory-Change As Structure-Change: Comments on the Sneed Formalism'.
[60] Ibid., pp. 190–1.
[61] Ibid., p. 191.
[62] Ibid., footnote 11, p. 198.
[63] Which appears in *The Essential Tension*, pp. 320–39. The following pages have also been incorporated, with minor changes, in a paper dealing with the distinction between the contexts of discovery and of justification. Cf. Siegel, 'Justification, Discovery, and the Naturalizing of Epistemology'. See also the discussion of Kuhn's essay in Frankel's 'Review of T.S. Kuhn, *The Essential Tension*.' Frankel makes good use of the distinction between *accepting* a theory (as true) and *pursuing* a theory (as plausible, or worthy of further consideration). This distinction is in turn developed nicely in Kordig, 'Discovery and Justification'.
[64] Reichenbach, *Experience and Prediction*, chapter 1. An excellent summary of Reichenbach's distinction is given in Scheffler, *Science and Subjectivity*, pp. 67–73. A discussion of the current status of the distinction is presented in my 'Justification, Discovery, and the Naturalizing of Epistemology'.
[65] *The Essential Tension*, p. 327.
[66] Ibid.
[67] Ibid.
[68] Ibid., pp. 327–8.
[69] Ibid., p. 328.
[70] Ibid.
[71] Ibid.
[72] Ibid.
[73] But cf. Chapter 4 below, and references to Scheffler and Shapere within.
[74] *The Essential Tension*, p. 337.
[75] *The Structure of Scientific Revolutions*, p. 149.
[76] *The Essential Tension*, pp. 338–9.
[77] Kuhn, 'Discussion,' in Suppe, *The Structure of Scientific Theories*, p. 508. Cf. also Kuhn, 'Postscript-1969', pp. 205–6.
[78] Besides the responses considered thus far, it is appropriate to mention two others here. Kuhn's 'Second Thoughts On Paradigms,' originally published in Suppe, *The Structure of Scientific Theories* and reprinted in Kuhn, *The Essential Tension*, pp. 293–319, attempts to refine his conception of paradigm. While space forbids detailed consideration of the point here, I believe it can be shown that Kuhn's refinement of the paradigm concept in that paper

substantially reduces the support that concept originally gave to the incommensurability thesis, thus further weakening the already questionable status of the incommensurability thesis, discussed above. I have briefly argued for this point in my 'Rationality, Talking Dogs, and Forms of Life'. In addition, Jack W. Meiland, in his 'Kuhn, Scheffler, and Objectivity in Science,' attempts to construe Kuhn as a proponent of objectivity in science. Consideration will be given to Meiland's paper in the following chapter.

79 Scheffler, 'Vision and Revolution: A Postscript on Kuhn', p. 369.

NOTES TO CHAPTER 4

1 Meiland, 'Kuhn, Scheffler, and Objectivity in Science.' All references to Meiland in this chapter are to this article unless otherwise noted.

2 Ibid., p. 179.

3 Ibid.

4 Scheffler, *Science and Subjectivity*, p. 89; cited in Meiland, p. 179. Kuhn puts the point in various ways: when "two scientific schools disagree . . . they will inevitably talk through each other when debating the relative merits of their respective paradigms" (Kuhn, *Structure*, p. 109); that paradigm debate is like "communication across the revolutionary divide" (ibid., p. 149); indeed, Kuhn even says "a decision of that kind [to accept a new paradigm] can only be made on faith" (ibid., p. 158).

5 Kuhn, ibid, pp. 52–3, cited in Meiland, p. 181.

6 Meiland, p. 186.

7 Ibid.

8 Ibid., p. 183.

9 Kuhn, *Structure*, p. 77; cited in Meiland, p. 183.

10 Meiland, p. 183.

11 Ibid.

12 There are other problems with this suggestion: 1) Such a criterion of evaluation would favor more cautious paradigms which did not try to solve difficult problems; 2) Paradigms would be "disconnected" – that is to say, a relatively successful paradigm of botany would be judged superior to a less successful paradigm of physics; and 3) It is not clear why such a criterion of evaluation would not be counted, on Meiland's terms, as a standard of SIS objectivity. We will return to this problem as it arises for Laudan in Chapter 6.

13 Kuhn, *Structure*, p. 158.

14 Meiland, p. 183.

15 Meiland, p. 184, citing Scheffler, *Science and Subjectivity*, pp. 85–6.

16 Scheffler, *Science and Subjectivity*, p. 83.

17 Kuhn, *Structure*, pp. 109–110, cited in Scheffler, *Science and Subjectivity*, p. 83.

18 Scheffler ibid., p. 84.

19 Meiland, p. 187.

20 Ibid., p. 179.

21 Ibid., p. 187.

22 Ibid., p. 179.

23 Kuhn, 'Reflections On My Critics,' p. 265; cited in Meiland, p. 185.

24 For instance, the quote above indicates the latter alternative, while the following passage from Meiland suggests the former: "For I think that what Scheffler is deploring, in this talk about reality being made rather than discovered, is that Kuhnian-style views seem

to rule out truth as a criterion of scientific progress." Meiland, p. 185.
[25] Cf. Scheffler, *Science and Subjectivity*, Chapter 5, esp. pp. 121–123.
[26] Scheffler, 'Vision and Revolution,' p. 369.
[27] Meiland, p. 187.
[28] Meiland, p. 184, citing Scheffler, *Science and Subjectivity*, pp. 85–6.
[29] Meiland, p. 187.
[30] Ibid., p. 179.
[31] Ibid., p. 187.
[32] Doppelt, 'Kuhn's Epistemological Relativism: An Interpretation and Defense.' All references to Doppelt in this chapter are to this paper unless otherwise noted.
[33] Ibid., p. 39.
[34] Ibid., p. 41.
[35] Ibid., p. 42–43.
[36] Ibid., p. 43.
[37] Ibid.
[38] Ibid., emphasis Doppelt's.
[39] Ibid., p. 44.
[40] Ibid., p. 45.
[41] One of Doppelt's major pre-occupations in his paper is to defend his reconstruction of Kuhn from Scheffler's and Shapere's (what Doppelt labels the "neo-positivist" interpretation) criticisms of Kuhn. While I believe a case could be made that Doppelt does not succeed in escaping the telling criticisms of Scheffler and Shapere, I shall not attempt to make such a case here. I forswear such a project because the more direct criticisms of Doppelt's view developed below are sufficient to render Doppelt's reconstruction of Kuhn implausible.
[42] Doppelt, p. 49.
[43] Ibid., p. 50.
[44] Ibid.
[45] Ibid.
[46] It is worth noting at this point that, far from rebutting the Scheffler-Shapere critique of incommensurability, Doppelt's position that proponents of incommensurable paradigms can coherently debate the merits of their respective paradigms actually grants these critics of Kuhn their main point – namely, that such meaningful debate is possible.
[47] Ibid., p. 41.
[48] Ibid., p. 50.
[49] Similarly with Einstein: Doppelt argues (pp. 81–82, Note 17) that Einstein's re-introduction of the need to explain gravitational forces, after Newtonian science had accepted the view that gravity did not need to be explained, demonstrates a powerful sort of incommensurability of "ultimate" explanatory adequacy. This sort of incommensurability, apparently, is "irresolvable by scientific argument" (p. 50) according to Doppelt. Yet it is clear that Einstein's "re-legitimation" of the problem of explaining gravity was not arbitrary or irrational. There were clear scientific reasons for taking up anew the problem of explaining gravity. Given the existence of such reasons, it is inappropriate to label such a difference "incommensurable."
[50] Ibid., p. 50, emphases added.
[51] Ibid., p. 53.
[52] Ibid., p. 53.

[53] Ibid., p. 39.
[54] Ibid.
[55] Ibid., p. 56.
[56] Ibid., pp. 52, 54, 56.
[57] Ibid., p. 56.
[58] Ibid.
[59] Ibid., p. 80, note 8.
[60] I note in passing that Doppelt's discussion of "compelling" (as opposed to "good") reasons is no advance over Kuhn's "no algorithm for theory choice" theme (cf. Chapter 3); and that Doppelt's rejection of "in principle" incommensurability in favor of a weaker version makes irrelevant his discussion of the *reductio* argument against relativism, for it is at the relativism which follows from the stronger version of incommensurability that the *reductio* is aimed.
[61] Doppelt, 'Relativism and the Reticulational Model of Scientific Rationality', p. 225. Cf. also Doppelt, 'The Philosophical Requirements for an Adequate Conception of Scientific Rationality'; Doppelt, 'Relativism and Recent Pragmatic Conceptions of Scientific Rationality'.
[62] Doppelt is concerned to relate moderate relativism to the problem of the rationality of science. While full consideration of this problem is beyond the bounds of the present effort, I should note that it is unclear why we should regard theory choice in the circumstance of under-determination as *rational*. For in that circumstance, while it is true that there are some reasons for the theory chosen, *ex hypothesi* those reasons do not favor the chosen theory *over its alternative(s)*. In such cases, I should think, rationality counsels not under-determined choice, but a postponement of choice until more powerful reasons for one alternative or another are at hand (or, if choice is forced, that such choice not be regarded as rationally justified). It is not clear (to me at least) why Doppelt regards under-determined choice as rational or as rationally justified. (Cf. my 'What Is the Question Concerning the Rationality of Science?' for further discussion.)
[63] It must, in fairness to Doppelt, be pointed out that his primary concern is to demonstrate the relevance of moderate relativism to the problem of the rationality of science. I do not claim to have settled anything about that alleged relevance here; nor have I offered any criticism of Doppelt's view of the problem of the rationality of science. I agree with Doppelt that the questions of rationality and relativism are tightly intertwined. (Cf. Chapter 8, below.) So, insofar as Doppelt is mainly concerned to utilize moderate relativism in the course of an argument concerning the rationality of science, I have said little about that main concern. My aim in the discussion of Doppelt's work here is simply that of showing that that work does not advance the prospects of a powerful epistemological relativism.

NOTES TO CHAPTER 5

[1] Brown, *Perception, Theory and Commitment*. All references to Brown in this chapter are to this book unless otherwise noted.
[2] Ibid., p. 11.
[3] Ibid., p. 10.
[4] Ibid., p. 26.

[5] Ibid., p. 37.
[6] Ibid., p. 29.
[7] I have argued against the new philosophy of science's tendency to conflate the attempt to understand the empistemic status of scientific claims with the attempt to understand the dynamics of the scientific enterprise as a whole in my 'Justification, Discovery, and the Naturalizing of Epistemology.'
[8] Actually Brown suggest that he is not criticizing the logical empiricists for their failure to attend to actual scientific theories and research, but only seeking to explain that failure by tracing it to the "empiricist assumption that all scientific knowledge consists of generalizations from experience" (30). However, since Brown takes the assumption to be a faulty one, his discussion does seem to be a criticism. We shall discuss the possibility of criticizing assumptions and presuppositions, given the new epistemology of science, presently.
[9] Brown, pp. 35–36.
[10] Ibid., p. 36.
[11] In fact it should be kept in mind that Goodman's work on induction is in large part an attempt to solve the problem of distinguishing accidental from lawlike generalizations. Cf. Scheffler, *The Anatomy of Inquiry*, pp. 295–314.
[12] Brown argues that older laws (e.g. Galileo's) are not strictly speaking deducible from newer (e.g. Newton's) laws. This is clearly right. We may all agree with Brown that, typically, a new law says something significant that the old law does not, and, from the point of view of understanding nature, indicates some weakness in the old law. To use Brown's example, for Galileo falling bodies fall with constant acceleration, while for Newton, acceleration increases steadily (in fact, is inversely proportional to the square of the distance from the body to the earth's center). Nevertheless, there is a clear sense in which the old law constitutes a "limiting case" of the newer law. In the example here considered, when the body's fall is relatively short, the increase in acceleration is negligible, and so the two laws predict (approximately) the same fall time. We can explain the success of Galileo's law for small distances, and its failure for large distances, by noting that that law's account of acceleration is mistaken in just such a way that the mistake is neglible for small distances, but significant for larger distances. Thus we can agree with Brown that the old law is not entailed by the new law, but there is still a clear sense in which the old law constitutes a limiting case of the new law, and in which the new law provides an explanation for the old law's success (in a limited domain) and for its failure. Cf. Kordig, 'Scientific Transitions, Meaning Invariance, and Derivability.'
[13] Brown, p. 66.
[14] Ibid., p. 63. This despite the fact that Brown cites a passage from Hempel which claims that Galileo's and Newton's laws are inconsistent (62).
[15] Ibid., p. 63.
[16] Ibid., p. 64. Emphasis mine.
[17] Cf. Scheffler, *Science and Subjectivity*, pp. 121–124. I shall return to this theme below.
[18] Brown, p. 64.
[19] Cf. Kuhn, *The Structure of Scientific Revolutions*, p. 169. Kuhn puts the point in terms of problem-solving ability, but it is clear that he is committed to some cumulativity across revolutions, though not cumulativity of theoretical propositions. The same is true of logical empiricism.
[20] Brown, p. 87.
[21] Ibid., p. 100.

[22] Ibid., p. 101.
[23] Ibid., p. 105.
[24] Ibid. Note the resemblance between Brown's paradigmatic propositions and Lakatos' "hard core." Cf. Lakatos, 'Falsification and the Methodology of Scientific Research Programmes.'
[25] Brown, p. 105.
[26] Ibid., p. 106.
[27] Ibid.
[28] Brown does make a move in this direction, by noting that "the reality that exists independently of our knowledge" (108) also has a hand in determining the fate of research programs, so that the presuppositional framework is not the sole determiner of that fate. But since the only access to this independent world "is through the creation of theories and the process of theory-directed research" (108), it is difficult to see how independent reality actually acts as a constraint on research in such a way as to limit the power of presuppositions to structure and determine the results of research – especially since Brown equates that independent reality, the world, with the scientist's "meaning structure" (111), which on Brown's account is clearly presupposition-bound.
[29] Ibid., p. 111.
[30] Ibid., pp. 115–116.
[31] Ibid., p. 118.
[32] Ibid.
[33] Ibid., p. 120.
[34] Cf. Scheffler, *Science and Subjectivity*, esp. Chapters 2 and 3.
[35] These points will be amplified below. Here we might note that Brown often appeals to criteria of justification when discussing cases of conceptual change (e.g., pp. 112, 117), without realizing that it is propositions, not concepts, that are being justified. This is especially clear, e.g., of his discussion of Galileo's re-analysis of "natural place" (117).
[36] Brown, p. 120.
[37] Brown does offer an account of the epistemic features of the new image of science; these will be considered presently.
[38] Ibid., p. 130.
[39] I have argued for this view in some detail in 'Justification, Discovery, and the Naturalizing of Epistemology.'
[40] Brown, p. 130.
[41] Ibid. pp. 129–130.
[42] For a clear discussion of the distinctions between discovery, plausibility, and justification, cf. Kordig, 'Discovery and Justification.' A helpful discussion of the methodological importance of distinguishing from among various cognitive stances one can take toward theories is that of Wykstra, 'Toward a Historical Meta-Method for Assessing Normative Methodologies: Rationality, Serendipity, and the Robinson Crusoe Fallacy,' esp. pp. 216–219.
[43] Cf. Thagard, 'Against Evolutionary Epistemology,' esp. pp. 188–189; Nickles, 'Scientific Problems: Three Empiricist Models,' esp. pp. 8–9; and Gutting, 'A Defense of the Logic of Discovery' and 'Science as Discovery.'
[44] 'Justification, Discovery, and the Naturalizing of Epistemology.'
[45] Cf. Nickles, 'Scientific Problems: Three Empiricist Models.'
[46] Brown, p. 133.

[47] Ibid.
[48] Ibid., p. 134.
[49] Ibid.
[50] Ibid., pp. 134–139.
[51] Ibid., p. 138.
[52] Ibid., p. 139–140.
[53] Ibid., p. 140.
[54] Ibid., p. 141.
[55] Ibid., p. 142.
[56] Ibid., p. 144.
[57] Ibid.
[58] Cf. my 'Justification, Discovery, and the Naturalizing of Epistemology.'
[59] Brown, p. 147.
[60] Cf. Nickles, 'Scientific Problems: Three Empiricist Models.'
[61] Brown, p. 147.
[62] Ibid., p. 148.
[63] Ibid., p. 149. This account of rational deliberation based on Aristotle's man of practical wisdom is further developed in Brown, 'On Being Rational,' esp. pp. 245–247.
[64] Brown, p. 149.
[65] Brown, 'On Being Rational,' p. 245.
[66] Cf. Brown, pp. 145; also Brown, 'On Being Rational.' As argued earlier, Brown's analysis of standard epistemology's conception of knowledge founders on the confusion, noted above, between truth and certainty (or infallibility). To say that, for a proposition to be known it must be true, is not to say that for it to be known it must be certain. I shall say a bit more about this below.
[67] Brown, p. 151.
[68] Though, as noted above, it is not the case that this separates his analysis of scientific knowledge from that of the logical empiricists, since they too held scientific knowledge to be fallible. Brown argues the contrary, but his argument confuses truth (which they do take to be a necessary condition for knowledge) with certainty or infallibility (which they do not).
[69] Ibid., p. 152.
[70] Ibid., emphasis added.
[71] This point is developed further in Chapter 8.
[72] Cf. Scheffler, *Science and Subjectivity*, pp. 121–124.
[73] Brown, pp. 151–152. Cf. also Chapter 1 above, pp. 9–10.
[74] Ibid., p. 153. As Brown justifiably complains ('Response to Siegel'), this way of putting the point is not quite right. As the next few lines in the text note, Brown distinguishes between two senses of truth: "truth_1" which Brown equates with "correct description of reality"; and "truth_2" which relativizes truth to current scientific knowledge. It is only the former, not the latter, which Brown gives up on. So strictly speaking, the text and quoted passage attendant to this footnote should read: ". . . it is impossible to 'discuss meaningfully the truth [i.e. truth_1] or falsity of actual scientific theories'." I am grateful to Brown for drawing my attention to this matter.
[75] Ibid.
[76] Ibid.
[77] Ibid., p. 154.

78 Ibid., p. 100.
79 Ibid., p. 155.
80 Recall that Brown's analysis is meant to apply to the philosophy of science as well as to science itself. Cf. ibid., pp. 10; 127; 166.
81 Cf. ibid., e.g. pp. 26, 30, 36.
82 Ibid., p. 31.
83 The interested reader may wish to consult Brown's 'Response to Siegel' for further discussion of the issues just treated. Unfortunately, detailed discussion of Brown's response to the analysis of his book given above would considerably extend an already overly long chapter, and while such discussion would be highly relevant in the context of a study of the new philosophy of science, it is of marginal interest here, in that Brown's response does not specifically address his defense of relativism.
84 An additional defender of Kuhn with respect to the issues we have been considering is Wolfgang Stegmüller, who in his *The Structure and Dynamics of Theories* presents a reconstruction of Kuhn's position which, Stegmüller claims, defends Kuhn from his critics (who Stegmüller labels the "critical rationalists"). Since some philosophers have acclaimed Stegmüller's reconstruction of Kuhn (for example, Kockelmans in his review of Stegmüller's book), it is worthwhile here to point out that a positive assessment of Stegmüller's reconstruction of Kuhn is problematic. While a comprehensive treatment of Stegmüller's treatise would take us far afield, it is possible to outline briefly the implications of Stegmüller's reconstrual of Kuhn's position for the issues presently being considered. These can be summed up as follows: Stegmüller claims that his account of scientific theories clarifies and supports Kuhn's view of the dynamics of theory change. However, concerning the issues we have been concerned with thus far – incommensurability, incomparability, relativism, rationality, and objectivity – Stegmüller's reconstrual, rather than defending Kuhn from his critics, actually grants Kuhn's critics their main points. The features of Kuhn's original presentation of theory choice, which his critics deplored – for example, the incommensurability of competing paradigms, the non-reducibility of old paradigms to their replacements, and the relativism implied by these positions – are not features of Stegmüller's reconstrual. On the contrary, according to that reconstrual, normal science is not irrational (p. 158); competing paradigms are not incommensurable (pp. 37, 245); progress during theory change can be objectively measured (p. 242); old paradigms are reducible to new ones (p. 216); a philosophically objectionable relativism is not maintained (pp. 154–7); and the criteria of evaluation of competing paradigms are not paradigm-bound (p. 154). In short, virtually all the aspects of Kuhn's views that the "critical rationalists" deplore are not aspects of Stegmüller's reconstrual of Kuhn. Stegmüller's reconstrual does not rebut Kuhn's critics; rather, it concedes their main points, and fails to clarify or improve Kuhn's position concerning incommensurability, comparability, relativism, and the objectivity of scientific theory choice.

As indicated, this brief treatment of Stegmüller is hardly comprehensive, but deals only with Stegmüller's reconstrual as it concerns the issues being considered in this volume. The reader interested in a more detailed discussion of Stegmüller's overall view is directed to his book, and to the review of that book by Joseph J. Kockelmans. Kuhn's attitude towards Stegmüller's reconstrual is given in his 'Theory-Change as Structure-Change: Comments on the Sneed Formalism.' Stegmüller's further developments of his view are presented in his 'Structures and Dynamics of Theories: Some Reflections on J.D. Sneed and T.S. Kuhn,' and his 'Accidental ("Non-Substantial") Theory Change and Theory

Dislodgement: To What Extent Logic Can Contribute to a Better Understanding of Certain Phenomena in the Dynamics of Theories.'

[85] One might expect, in a book like this, a discussion of Feyerabend's (views on) relativism. But Feyerabend's position precludes the possibility of there being any intellectual responsibility to offer such a discussion. So I am simply following his advice and exercising my "Dadaism" in refusing to consider his views here. The legitimacy, on his view, of such a refusal constitutes the cleanest objection to his view I am aware of, and exhibits the impotence of (Feyerabendian) relativism discussed in Chapter 1. Cf. Newton-Smith, *The Rationality of Science*, Chapter 6, esp. p. 127.

NOTES TO CHAPTER 6

[1] Toulmin, *Human Understanding*, Volume I. All references to Toulmin in this chapter are to this volume unless otherwise noted.

[2] Laudan, *Progress and Its Problems*. All references to Laudan in this chapter are to this volume unless otherwise noted.

[3] Cf. Toulmin, pp. 98–117; Laudan, pp. 73–76 and 133–136.

[4] Toulmin, p. x.

[5] Here we immediately face the problem noted for Brown in Chapter 5: is it *concepts* that demand rational justification, or *claims* (hypotheses, theories, etc.) framed in terms of concepts that need to be rationally justified? This is an ambiguity running throughout Toulmin's book – Toulmin typically talks of concepts having rational warrant, when the epistemic object having warrant is a claim made in terms of some concept. We shall return to this problem in due course.

[6] Toulmin, pp. 41–52.

[7] Ibid., p. 44.

[8] Ibid.

[9] Ibid., p. 61.

[10] Ibid.

[11] Ibid., p. 64.

[12] Ibid., p. 67.

[13] Ibid., p. 54, emphasis Toulmin's.

[14] Ibid., pp. 122 ff.

[15] Ibid., pp. 146–9. It must be pointed out here that problems evolve as well as concepts. We shall return to this point below.

[16] Ibid., pp. 150–1. Note here the similarity to Laudan's distinction between empirical and conceptual problems.

[17] Ibid., p. 152.

[18] Ibid.

[19] Note the looming problem for Toulmin's account, namely, the constancy of explanatory ideals. What happens when these change? We will take up this question below.

[20] Ibid., p. 166.

[21] Ibid., p. 168.

[22] Ibid., p. 225.

[23] Ibid., p. 237.

[24] Ibid., p. 239.

[25] Ibid.
[26] Ibid., p. 241.
[27] While some of these difficulties are not directly related to Toulmin's specific claims concerning problem-solving ability, they do speak directly to various aspects of his overall analysis of rationality, upon which the problem-solving claims rest, and so are relevant to the evaluation of those more specific claims.
[28] Ibid., p. 152.
[29] This point is well made by E. McMullin, 'Logicality and Rationality: Comments on Toulmin's Theory of Science', esp. pp. 427–8, and note 7, p. 430. Cf. also I.C. Jarvie, '*Human Understanding I* by Stephen Toulmin,' p. 93.
[30] Toulmin, p. 61.
[31] Here, and throughout this paragraph, I am indebted to Carl R. Kordig's 'Review of *Human Understanding, Volume I*, by Stephen Toulmin', esp. pp. 536–8.
[32] Cf. Kordig, ibid., p. 538. As Kordig points out, there are other problems with Toulmin's analysis of rationality in terms of furthering disciplinary goals. Toulmin also succumbs to a difficulty opposite to the one sketched in the text, in that his account of certain sorts of disciplinary judgment (concerning "cloudy cases") is relativistic. This point will be developed below.
[33] Toulmin, p. 64.
[34] Here again I am indebted to McMullin, 'Logicality and Rationality', pp. 418–19 and 423–6. A similar point is made in my 'Justification, Discovery, and the Naturalizing of Epistemology', pp. 308–9.
[35] The points made in this paragraph concerning the possibility of logical relations across rival conceptual systems were made by Professors Israel Scheffler and Carl R. Kordig in triangular correspondence with me during 1978–9. While entirely theirs, these points seemed to me important enough to be more widely known, and so, with their kind permission, I present them here.
[36] Toulmin, p. 241.
[37] Ibid., pp. 234–6.
[38] Ibid., p. 237.
[39] Ibid., p. 238.
[40] Ibid., p. 239.
[41] Ibid.
[42] Ibid., p. 240.
[43] Additional difficulties with Toulmin's reliance on the analogy between science and the law are forcefully noted in McMullin, 'Logicality and Rationality', p. 429.
[44] Toulmin, pp. 239–41.
[45] Ibid., p. 241.
[46] While I have not argued the case here, I believe that Toulmin's account of choice in cloudy cases can be made to work only by recasting it in absolutist form. For related discussion, cf. Kordig's 'Review', op. cit., p. 536, and Briskman, 'Toulmin's Evolutionary Epistemology', pp. 168–9.
[47] Cf. Briskman, ibid., pp. 166–8.
[48] E. McMullin, in his review of Laudan's book ('Discussion Review': Laudan's *Progress and Its Problems*'), points out the "obvious affinity" between Toulmin's and Laudan's analyses (pp. 625–6) that I am attempting to document and explore here. I should note that my discussion of Laudan is narrow, focusing mainly on his analysis of rationality.

There is much of interest and merit in Laudan's book that I cannot consider here. For a more detailed analysis of Laudan's account of the rationality of science and its relation to relativism, cf. Gerald Doppelt's good analyses, 'Laudan's Pragmatic Alternative to Positivist and Historicist Theories of Science', and 'Relativism and Recent Pragmatic Conceptions of Scientific Rationality'.

[49] Laudan, pp. 11, 68.

[50] Ibid., p. 68, emphasis Laudan's.

[51] Ibid., p. 125, emphasis Laudan's. See also p. 109. But see also McMullin's argument that, in the final analysis, Laudan falls back on the traditional view that rationality is primary and progressiveness derivative, in McMullin's 'Review', *op. cit.*, p. 632. While I have not developed the point here, I believe McMullin is correct that Laudan's reliance on "pre-analytic intuitions" of rationality is both problematic in itself and also contrary to the thrust of Laudan's approach, which is to make rationality a function of problem-solving effectiveness and progressiveness rather than the reverse.

[52] Laudan, p. 125, emphasis Laudan's. It is perhaps worth noting that Kuhn, too, construes science such that science is not a truth-seeking activity – though not in the context of developing an account of the rationality of science, as do Toulmin and Laudan. For discussion of this aspect of Kuhn, cf. Shimony, 'Comments on Two Epistemological Theses of Thomas Kuhn', and Chapter 4, A, above.

[53] McMullin, 'Review', p. 627.

[54] Laudan, p. 146.

[55] Ibid., emphasis Laudan's.

[56] Ibid., p. 144, emphasis Laudan's.

[57] A more detailed account of the Dalton/pre-Dalton case, especially with respect to the possibility of assessing the rationality of rival research traditions and paradigms, can be found in my 'On the Parallel Between Piagetian Cognitive Development and the History of Science', and in this volume, Chapter 4, B, regarding my exchange with Gerald Doppelt.

[58] This point is made in Kordig, 'Progress Requires Invariance.'

[59] Discussion of similar difficulties for Kuhn's attempt to account for progress in science across revolutions in terms of a paradigm's internal problem-solving ability can be found in Chapter 4, A, above.

[60] The relativistic tendency of Laudan's analysis is clearly developed in Doppelt, 'Laudan's Pragmatic Alternative to Positivist and Historicist Theories of Science,' esp. pp. 264–267.

It is perhaps worth pointing out related difficulties with Laudan's discussion of standards of rationality. Laudan criticizes Scheffler's view that there are constant, time-independent standards of rational appraisal. (Scheffler speaks of a "constancy of logic and method, which unifies" different scientific periods.) Laudan criticizes this view on the grounds that "such components of rational appraisal as criteria of explanation, views about scientific testing, beliefs about the methods of inductive inference and the like have undergone enormous transformations." (p. 129). There are two points which need to be made here. First, despite Laudan's criticism of methodological constancy, he applauds his own account of rationality for accommodating "the more general, time-independent features of rational decision making" (p. 131). Problem-solving effectiveness is itself a methodological constant of precisely the sort Laudan seems to want to disallow. Second, Laudan's criticism of Scheffler is not quite fair, because Scheffler's position is entirely compatible with the sorts of "enormous transformations" Laudan notes. Scheffler's view is not that views concerning what counts as inductive support, proper test, satisfactory explanation and so on never change. It is rather that methodological constancy involves not specific construals of various methodological criteria, but a general commitment to rational procedures and standards without which the entire enterprise of science would be impossible. Cf. here cogent relevant discussion in A. Musgrave, 'Problems with Progress,' esp. pp. 451–2 and 456–9.

[61] This last point, that problem-solving effectiveness cannot serve as a tradition-neutral criterion for evaluating research traditions, is well made in G. Gutting's review of Laudan, 'Review of Larry Laudan, *Progress and Its Problems*', p. 99; and also in Kordig, 'Progress Requires Invariance'. See also the problem with Laudan's apparatus for assessing a theory's problem-solving effectiveness raised (briefly) in Glymour, *Theory and Evidence*, pp. 100–1. Criticism of Toulmin analogous to that raised in the above paragraph against Laudan is made in Kordig's 'Review', *op. cit.*

[62] Toulmin, pp. 168–173.

[63] Kordig, 'Review', p. 539.

[64] Toulmin seems to accept this point (p. 173), without realizing that it contradicts his earlier (pp. 169–171) denial of it.

[65] Laudan, p. 125, emphasis Laudan's.

[66] Ibid., emphasis Laudan's. Laudan continues this passage by despairing of efforts to relate rationality with approximations to the truth or probable truth, but for present purposes we may ignore this aspect of Laudan's discussion.

[67] Cf. Scheffler, *Science and Subjectivity*, chap 5, esp. pp. 121–4. Though Scheffler's argument here concerns the problem of reference, his distinction between the methods of establishing truth and the import or purport of such establishment is parallel to the distinction I am drawing in the text. A similar point with respect to simplicity is made by Nelson Goodman, *Problems and Projects*, p. 280. Cf. also Firth, 'Epistemic Merit, Intrinsic and Instrumental', esp. p. 19.

[68] Laudan's later remarks concerning the need for a truth-independent account of rationality fail to improve his original argument. In his 'Views of Progress: Separating the Pilgrims from the Rakes', pp. 277–8, Laudan reformulates the argument that the truth of a theory cannot provide us with grounds for rationally accepting a theory, so that truth cannot be appealed to as evidence for rational belief, and so that the rationality of science cannot be established in terms of science's achieving (approximately) true theories. But this reformulation fails to carry Laudan's view forward, because it again reverses the relation between rationality and truth.

[69] It is perhaps worth pointing out that Laudan grants (p. 82) that "one can conceive that a research tradition might be true," but holds that the success of a successful (i.e., problem-solving effective) tradition does not reflect on that tradition's truth. But, as we have seen, such success does count for Laudan as grounds for rational belief in a tradition (p. 124) – that is, for Laudan it is rational to accept those research traditions which are maximally progressive and problem-solving effective. Granting that point, and given the argument just sketched in the text, it is not clear why Laudan is not committed to the conclusion that progressiveness and problem-solving effectiveness, in providing good reason for accepting research traditions, provides grounds for the truth of those traditions as well.

[70] Musgrave, 'Problems with Progress', p. 460.
[71] Musgrave, 'Problems with Progress', p. 460; Gutting, 'Review', p. 96.
[72] McMullin, 'Review', pp. 633–5; Gutting, 'Review', p. 97.
[73] Gutting, 'Review', pp. 97–8. Cf. also my 'What is the Question Concerning the Rationality of Science?'.
[74] Laudan, p. 225.
[75] Musgrave, 'Problems with Progress', pp. 460–1; Gutting, 'Review', p. 98.
[76] Several writers besides the ones just cited have pointed to this problem for Laudan. Cf. also, for example, the papers by D.L. Hull, A. Lugg, and R.E. Butts in the review symposium on Laudan, in *Philosophy of the Social Sciences* of December 1979 (Hull, 'Laudan's Progress and Its Problems'; Lugg, 'Laudan and the Problem-Solving Approach to Scientific Progress and Rationality'; and Butts, 'Scientific Progress: The Laudan Manifesto', esp. pp. 462–4 and 473–4.

Laudan has attempted, unsuccessfully in my view, to deflect this criticism. Cf. his 'Problems, Truth, and Consistency,' and references therein.
[77] The suggestion here, that we explain a theory's success (understood as problem-solving effectiveness) in terms of truth, is a part of the 'Putnam-Boyd thesis' that realism is an empirical hypothesis which explains the success of science. Cf. Putnam, *Meaning and the Moral Sciences*, and Skagestad, 'Pragmatic Realism: The Peircean Argument Reexamined', esp. pp. 527–9.

NOTES TO CHAPTER 7

[1] *Structure of Appearance*, p. xiii. Henceforth referred to as *SA*.
[2] Israel Scheffler, 'The Wonderful Worlds of Goodman,' p. 201.
[3] Thus Goodman's ontological relativism differs markedly from Quine's, since for Quine there *is* a privileged system, namely that of physical theory, and so a basic (though relative) ontology, namely that of current physical theory. A helpful short introduction to Goodman's position is that of Geoffrey Hellman, 'Introduction.'
[4] *Structure of Appearance*, pp. 99–104.
[5] 'The Way the World Is' (henceforth *WWI*).
[6] *WWI*, pp. 30–31, emphasis in original.
[7] Goodman does not claim (rightly, I think) that truth is itself a criterion by which descriptions are judged. Rather, truth for Goodman is the upshot of a description's meeting various other criteria. Goodman does not offer a special analysis of criteria of truth, so far as I am aware (though he does discuss tests of truth in *Ways of Worldmaking*, chapter 7). Indeed, if I understand him, there can be no such criteria (*of* truth), but only criteria of, e.g., credibility, or simplicity, such that, if a description met those criteria, that fact would tend to point to that description's being true.
[8] Henceforth *WW*. Page reference in the text are to this work.
[9] In addition to *WW*, cf. also Goodman, *Languages of Art*. For a helpful discussion of the connection between "constructionalism" and relativism, cf. Chris Swoyer, 'True For'.

In a longer version of this chapter, 'Relativism, Realism and Rightness: Notes on Goodmanian Worldmaking', I discuss the way that Goodman's constructionalism leads naturally to the view he labels "irrealism", which is defended at length in *WW*. In the present version I instead proceed directly to the consideration of Goodman's discussion of "criteria of rightness", the status of which is crucial to my argument concerning Goodmanian relativism.

[10] Goodman, 'Replies,' p. 159.

[11] It is perhaps worth pointing out the affinities between Goodman's discussion of "fit" and that of Hilary Putnam in *Reason, Truth and History*. Putnam in that work embraces, and apparently regards as unproblematic, Goodman's analysis of fit (Cf. pp. 54–5; 64; 123). In fact, Putnam's general position, which he terms "internal realism", is remarkably close to Goodman's "irrealism." (In addition to the pages just cited, cf. pp. 52–4; 79; also Putnam, 'Realism and Reason'.)

[12] I am here challenging Goodman's account of justification in terms of mutual accommodation of theory and practice (which Rawls labels "reflective equilibrium."). Cf. Goodman, *Fact, Fiction and Forecast*, pp. 62–66. Space forbids my pressing the challenge here. I do so in another article, 'Justification by Balance' (unpublished).

Catherine Elgin, in her 'Goodman's Rigorous Relativism', suggests that I misinterpret Goodman on rightness and fit. She claims that I conclude that in Goodman's view a practice cannot be wrong. But this is false: I cite Goodman (*WW*, p. 128; 135–7) as allowing that practices can be wrong and can be improved. My point, rather, is that if Goodman is correct that practices can go wrong and be assessed and improved, then there must be criteria (other than practice itself) by which such assessment can be carried out; and that if there are such criteria, moreover, then rightness cannot be a matter of fit with practice but must rather be a matter of sanction by such criteria. Thus, if practice can be improved, go wrong, and be assessed, then rightness cannot be a matter of fit with practice. (In Elgin's example of logical inference on p. 40, rightness is a matter, not of fit, but of logical consistency.) Elgin also fails successfully to challenge my claim that, on Goodman's view, "anything can be right in that it can be seen to fit some practice, if we are only perverse enough, or liberal enough, in our consideration of practices." Her example of mathematicians being unwilling to reject the law of non-contradiction, for a mathematics which rejects that law (and so the criterion of logical consistency) would be "incapable of serving the cognitive goals of mathematics or the cognitive and practical goals of the various enterprises to which mathematics contributes" (p. 41), does not in the least defeat my claim. For if we had different goals then the standard ones Elgin refers to – if, for example, our goal was to increase frustration among mathematics students, because we thought that such frustration would enhance learning or character development – then we *should* be willing to reject the law of non-contradiction. Mathematical practice which failed to secure consistency, given this (perverse) goal, would on Goodman's view of rightness be right. Criticism of this goal, moreover, would require appeal to criteria by which practical goals could be assessed, thus again showing that rightness cannot be a matter of fit with practice. Thus Elgin's arguments fail to rescue Goodman's analysis of rightness in terms of fit with practice from the criticisms offered in the text.

[13] It must be noted that Goodman presents his discussion of rightness as tentative and preliminary. Consequently my remarks here should be taken as a call for further clarification, rather than as a final critique of a fully worked out position.

[14] Note that the relativity of truth held in *WW* is a shift from the view put forth in *WWI* that truth is not relative.

[15] The point is formulated slightly differently in its original presentation in 'Words, Works, Worlds', pp. 70–71.

[16] Cf. Putnam, *Reason, Truth and History*, p. 123. It follows that criteria of rightness are here taken to be *external* to the versions being evaluated. For the analogous issue with respect to Kuhn's claim that standards of paradigm-evaluation are *internal* to paradigms, cf. Chapter Three above.
[17] Cf. Hellman, 'Introduction', pp. xxxvi–xxxvii.
[18] As Catherine Elgin argues, 'Goodman's Rigorous Relativism', pp. 41–44. Israel Scheffler, in conversation and correspondence, has also offered this interpretation of my argument. I am grateful to Scheffler and to Elgin for their criticisms and suggestions on this matter, though I have not been persuaded by their view.
[19] There is an additional difficulty here, with underscores the problematic nature of Goodman's criterion of "fit." For suppose we could somehow overcome the problem just noted, and figure out a way for Goodman to claim that his meta-version is cognitively superior to or epistemically more worthy than rival meta-versions. Any argument in support of this claim would depend on the criteria his meta-version sets out, i.e., on his meta-version's "fit with practice." But (a) his meta-version in fact does *not* fit well, since our practice includes many assumptions of "absolutism" – we typically act, for example, as if there *is* a real world which exists independently of us and our versions, and that there are facts of the matter with respect to all sorts of questions and matters which might be found, but are not made, by us. Moreover (b) this just-mentioned fact illustrates the difficulty (noted earlier) stemming from the emphasis and weight Goodman's criteria put on fit with *practice*. For Goodman should want to hold, in arguing for the worthiness and preferability of his meta-version, that if his meta-version does not fit with our practice, it may nevertheless *still be right* or preferable to meta-versions which more closely fit our practice. Thus practice must itself admit of evaluation, in terms other than fit with itself, or else new practices and new views (such as Goodman's) could never turn out to be rationally preferable or superior to their ancestors.
[20] This last point is well made in Hellman, 'Introduction', p. xlii.
[21] It might be argued (as Elgin does, 'Goodman's Rigorous Relativism', p. 42) that I am being unfair to Goodman here, for it does not follow, from the fact that criteria must discriminate, that another criterion may not be equally good even though it discriminates differently. I believe that this conclusion *does* follow. For if a second criterion does discriminate equally as well as, but differently from, a first, we cannot say that the first (or the second) criterion successfully discriminates as originally intended at all. Consider again the spelling bee. (I borrow here from Elgin's example, p. 42.) If we are using the *American Heritage Dictionary* as our standard of correct spelling, then "honor" will meet the criterion, but not "honour", while if we use the *Oxford English Dictionary* that result will be reversed. Ask now: does the criterion "spells as the *American Heritage Dictionary* spells" constitute a criterion of correct spelling? It does, I claim, only if we can argue that spellings from that dictionary are in some way objectively preferable to spellings from the *O.E.D.* If we cannot so argue, then the criterion does not serve as a criterion of *correct* spelling. Rather, it serves as a criterion of something weaker: perhaps "preferred" or "*locally* correct" spelling. In short, if the criterion is genuinely to discriminate among particular objective features or properties, there cannot be alternative, equally legitimate criteria which discriminate differently. Conversely, if other criteria are regarded as equally legitimate, then we cannot regard the original criterion as discriminating among those particular objective features or properties. Elgin suggests that ". . . the capacity to serve as a criterion does not require the denial that there are alternatives, or even the denial that

our choice among acceptable alternatives is ultimately arbitrary. Nor does the recognition of such alternative ways of deciding correct spelling amount to the admission that . . . there is no genuine difference between correct and incorrect spelling." (p. 42) This I think is clearly false. The admission that the spellings of the *O.E.D.* are equally as legitimate as the spellings of the *A.H.D.* *does* amount to the admission that the *A.H.D.* fails to serve as a standard of *correct* spelling. It rather serves as a standard of "locally preferred" (or some such) spelling. Thus spelling bees are not concerned with *correct* spelling at all. On the other hand, to take seriously the idea that the student who spells "honor" with a "u" has objectively misspelled that word is to reject the *O.E.D.* as an alternative standard of objectively correct spelling.

Another way to put this point is as follows. One could argue that Goodman's criteria of rightness do discriminate between versions, namely between right and wrong ones. And in a sense they do. However, they do so only *impotently*. (Cf. the discussion of relativism's impotence in Chapter 1). For the discriminations offered by Goodman's criteria have no cognitive or epistemic force because in Goodman's meta-version they are of no more epistemic significance than rival, incompatible criteria. Thus there is no reason to prefer versions picked out as right by Goodman's criteria to other versions picked out as right by alternative criteria. So Goodman's criteria do discriminate, but only impotently. (This paragraph is also relevant to the charge that I have fallaciously reasoned from "some" to "all", considered above. Goodman says "some, but not all, versions and meta-versions are right". But the impotence of his criteria – that is, the fact that they are criteria only from the point of view of his own meta-version – commits him to "all . . .".)

22 'Replies,' p. 159. Cf. also *WW*, p. 107.

23 Hilary Putnam, 'Reflections on Goodman's *Ways of Worldmaking*.'

24 Ibid., p. 617.

25 Ibid., p. 618, citing *WW*, footnote, pp. 109–110. It is ironic that (as others have pointed out) it is not clear that Putnam's positive view (his "internal realism") is not itself relativistic, and so open to the same difficulty he here points out for Goodman. But this point cannot be pursued here.

26 Goodman, 'Credence, Credibility, Comprehension,' p. 619.

27 In his fuller reply to Putnam ('On Starmaking,' pp. 211–212), Goodman extends the just-cited quote, and concludes that "Relativity goes all the way up." (211) Here again Goodman needlessly invites the difficulties that plague GR_2. If relativity occurs at the meta-level, then his rigorous restraints on radical relativism dissolve.

28 I am grateful to Catherine Z. Elgin for reminding me of this.

29 E.g. Catherine Elgin, Israel Scheffler, and, in some passages, Goodman himself.

NOTES TO CHAPTER 8

1 Beach, 'The Paradox of Cognitive Relativism Revisited,' p. 3. Emphases in original. Beach forswears the task of deciding between the two positions in his paper, in which he seeks to demonstrate that the third alternative, relativism, is untenable.

2 Ibid.

3 Field, 'Realism and Relativism,' p. 565, emphasis Field's. Cf. Chapter 1 above.

4 Brown, 'For a Modest Historicism,' pp. 549–550. Cf. Chapter 4 above.

[5] As Putnam puts it, ". . . it is a presupposition of thought itself that some kind of objective 'rightness' exists." *Reason, Truth and History*, p. 124.
[6] Cf., e.g., Popper's discussion of "fallibilistic absolutism," in *The Open Society and Its Enemies*, pp. 377–8. An excellent introductory discussion of fallibilist epistemology which avoids certainty but maintains an absolutist conception of truth may be found in Israel Scheffler's *Conditions of Knowledge*, Chapter 2. Cf. also Scheffler's discussion of the avoidance of certainty in favor of (Russell's and) Goodman's notion of credibility in *Science and Subjectivity*, Chapter 5.
[7] Cf. my 'Relativism, Rationality, and Science Education' for further discussion of the distinction between relativism and pluralism drawn in the text.
[8] I speak here, of course, of philosophical, not psychological, motivation.
[9] And Elgin, Scheffler, et. al. Cf. Chapter 7.
[10] From now on I will, by "absolutism," be referring to the non-vulgar variety.
[11] This is argued for by Michael Krausz, 'Relativism and Foundationalism: Some Distinctions and Strategies.'
[12] Although much more clearly needs to be said concerning the nature and status of the criteria just mentioned. I shall return to this matter below.
[13] This is of course just another instance of what I called in Chapter 1 relativism's impotence.
[14] Cf., for example, Trigg, *Reason and Commitment*, Chapter 7, for a typical expression of this view. It is a view widely held in the literature.
[15] E.g. Popper, *The Open Society and Its Enemies*, Volume 2, pp. 230–231.
[16] And of course even if the obligation is not met, the relativist must still confront the anti-relativist arguments offered in earlier chapters. I hope, then, that the criticisms of relativism formulated here constitute an advance in our understanding of the relativism/absolutism debate, even in the absence of the development of a constructive alternative.

BIBLIOGRAPHY

Asquith, Peter D., and Giere, Ronald N. (eds.) (1980), *PSA 1980* (Proceedings of the 1980 Biennial Meeting of the Philosophy of Science Association), East Lansing, Michigan: Philosophy of Science Association.

Barker, Peter (1981), 'Uncle Ludwig's Book About Science'. *Proceedings of the Southwestern Philosophical Society*, Supplement to *Philosophical Topics*, pp. 71–78.

Beach, Edward (1984), 'The Paradox of Cognitive Relativism Revisited: A Reply to Jack W. Meiland'. *Metaphilosophy* **15**, 1–15.

Black, Max (1978), '*Lebensform* and *Sprachspiel* in Wittgenstein's Later Work'. In *Wittgenstein and His Impact on Contemporary Thought*. Proceeding's of the 2nd International Wittgenstein Symposium, Kirchberg, Austria. Wien: Hölder-Pichler-Tempsky.

Briskman, Larry (1974), 'Toulmin's Evolutionary Epistemology'. *The Philosophical Quarterly* **24**, 160–169.

Brandt, Richard B. (1979), *A Theory of the Good and the Right*. Oxford: Clarenden Press of Oxford University Press.

Brandt, Richard B. (1984), 'Relativism Refuted?' *The Monist* **67**, 297–307.

Brown, Harold I. (1977a), *Perception, Theory, and Commitment*. Chicago: Precedent Publishing, Inc.

Brown, Harold I. (1977b), 'For a Modest Historicism'. *The Monist* **60**, 540–555.

Brown, Harold I. (1978), 'On Being Rational'. *American Philosophical Quarterly* **15**, 241–248.

Brown, Harold I. (1983), 'Response to Siegel'. *Synthese* **56**, 91–105.

Burnyeat, M.F. (1976a), 'Protagoras and Self-Refutation in Plato's Theaetetus'. *The Philosophical Review* **LXXXV**, 172–195.

Burnyeat, M.F. (1976b), 'Protagoras and Self-Refutation in Later Greek Philosophy'. *The Philosophical Review* **LXXXV**, 44–69.

Butts, Robert E. (1979), 'Scientific Progress: The Laudan Manifesto'. *Philosophy of the Social Sciences* **9**, 475–483.

Davidson, Donald (1973), 'On the Very Idea of a Conceptual Scheme'. *Proceedings of the American Philosophical Association*, 1973–74, pp. 5–20. Reprinted in Meiland and Krausz, pp. 66–80.

Davidson, Donald, and Jaakko Hintikka (eds.) (1969), *Words and Objections: Essays on the Work of W.V. Quine*. Dordrecht: D. Reidel Publishing Company.

Doppelt, Gerald (1978), 'Kuhn's Epistemological Relativism: An Interpretation and Defense'. *Inquiry* **21**, 33–86. Reprinted in Meiland and Krausz, pp. 113–146.

Doppelt, Gerald (1980), 'A Reply to Siegel on Kuhnian Relativism'. *Inquiry* **23**, 117–123.

Doppelt, Gerald (1981), 'Laudan's Pragmatic Alternative to Positivist and Historicist Theories of Science'. *Inquiry* **24**, 253–271.

Doppelt, Gerald (1983), 'Relativism and Recent Pragmatic Conceptions of Scientific Rationality'. In Nicholas Rescher (ed.), *Scientific Explanation and Understanding*. Lanham, Maryland: University Press of America, pp. 107–142.

Doppelt, Gerald (1987a), 'Relativism and the Reticulational Model of Scientific Rationality'. *Synthese* (in press).

Doppelt, Gerald (1987b), 'The Philosophical Requirements for an Adequate Conception of Scientific Rationality'. *Philosophy of Science* (in press).

Elgin, Catherine Z. (1983), *With Reference to Reference*. Indianapolis, Indiana: Hackett Publishing Company.

Elgin, Catherine Z. (1984), 'Goodman's Rigorous Relativism'. *Journal of Thought* **19**, 36–45.

Field, Hartry (1982), 'Realism and Relativism'. *Journal of Philosophy* **LXXIX**, 553–567.

Firth, Roderick (1981), 'Epistemic Merit, Intrinsic and Instrumental'. *Proceedings and Addresses of the American Philosophical Association* **55**, 5–23.

Frankel, Henry (1978), 'Review of T.S. Kuhn, *The Essential Tension*'. *Philosophy of Science* **45**, 649–52.

Glymour, Clark (1980), *Theory and Evidence*. Princeton: Princeton University Press.

Goodman, Nelson (1968), *Languages of Art*. Indianapolis: Bobbs-Merrill.

Goodman, Nelson (1972a), 'The Way The World Is'. Reprinted in Goodman, *Problems and Projects*, pp. 24–32. Originally published in *Review of Metaphysics* **14**, (1960), 48–56.

Goodman, Nelson (1972b), *Problems and Projects*. Indianapolis: Bobbs-Merrill.

Goodman, Nelson (1975), 'Words, Works, Worlds'. *Erkenntnis* **9**, 57–73.

Goodman, Nelson (1977), *The Structure of Appearance*. Third Edition. Dordrecht: D. Reidel Publishing Co. First published in 1951.

Goodman, Nelson (1978a), 'Replies'. *Erkenntis* **12**, 153–179.

Goodman, Nelson (1978b), *Ways of Worldmaking*. Indianapolis: Hackett Publishing Co., Inc.

Goodman, Nelson (1979), 'Credence, Credibility, Comprehension'. *Journal of Philosophy* **LXXVI**, 618–619.

Goodman, Nelson (1980), 'On Starmaking'. *Synthese* **45**, 211–215.

Goodman, Nelson (1983), *Fact, Fiction and Forecast*. 4th edition. Cambridge: Harvard University Press.

Grandy, Richard (ed.) (1973), *Theories and Observation in Science*. Englewood Cliffs, New Jersey: Prentice-Hall.

Gutting, Gary (1973), 'A Defense of the Logic of Discovery'. *Philosophical Forum* **4**, 384–405.

Gutting, Gary (ed.) (1980a), *Paradigms and Revolutions*. Notre Dame: Univ. of Notre Dame Press.

Gutting, Gary (1980b), 'Science As Discovery'. *Review Internationale de Philosophie*, pp. 26–48.

Gutting, Gary (1980c), 'Review of Larry Laudan, *Progress and Its Problems*'. *Erkenntnis* **15**, 91–103.

Harman, Gilbert (1975), 'Moral Relativism Defended'. *Philosophical Review* **84**, 3–22. Reprinted in Meiland and Krausz, pp. 189–204.

Harman, Gilbert (1977), *The Nature of Morality*. New York: Oxford University Press.

Hellman, Geoffrey (1977), 'Introduction'. In Goodman, *The Structure of Appearance*, pp. xix-xlvii.

Hinman, Lawrence (1983), 'Can a Form of Life Be Wrong?' *Philosophy* **58**, 339–351.

Hollis, Martin, and Steven Lukes (eds.) (1982), *Rationality and Relativism*. Cambridge: MIT Press.

Hull, David L. (1979), 'Laudan's Progress and Its Problems'. *Philosophy of the Social Sciences* **9**, 457–465.

Jarvie, I.C. (1975), '*Human Understanding I* by Stephen Toulmin'. *Philosophy of the Social Sciences* **5**, 91–94.

Jordan, James N. (1971), 'Protagoras and Relativism: Criticisms Bad and Good'. *Southwestern Journal of Philosophy* **2**, 7–29.

Kitcher, Philip (1978), 'Theories, Theorists, and Theoretical Change'. *Philosophical Review* **87**, 519–47.

Kockelmans, Joseph J. (1976), 'Review of W. Stegmüller, *The Structure and Dynamics of Theories*'. *Philosophy of Science* **43**, 293–7.

Kordig, Carl R. (1971a), *The Justification of Scientific Change*. Dordrecht: D. Reidel Publishing Company.

Kordig, Carl R. (1971b), Scientific Transitions, Meaning Invariance, and Derivability'. *Southern Journal of Philosophy*, pp. 119–125.

Kordig, Carl R. (1976), 'Review of *Human Understanding, Volume I*, by Stephen Toulmin'. *Philosophia* **6**, 533–40.

Kordig, Carl R. (1978), 'Discovery and Justification'. *Philosophy of Science* **45**, 110–17.

Kordig, Carl R. (1980), 'Progress Requires Invariance'. *Philosophy of Science* **47**, 141.

Kuhn, Thomas S. (1957), *The Copernican Revolution: Planetary Astronomy in the Development of Western Thought*. Cambridge: Harvard University Press.

Kuhn, Thomas S. (1969), 'Postscript–1969'. In Kuhn, *The Structure of Scientific Revolutions*, pp. 174–210.

Kuhn, Thomas S. (1970a), *The Structure of Scientific Revolutions*. First edition, 1962; second edition, enlarged. Chicago: University of Chicago Press.

Kuhn Thomas S. (1970b), 'Reflections on my Critics'. Included in Lakatos and Musgrave, pp. 231–78.

Kuhn, Thomas S. (1971), 'Notes on Lakatos'. In R. Buck and R. Cohen (eds.): *Boston Studies in the Philosophy of Science*, Vol. 8. Dordrecht: D. Reidel Publishing Company.

Kuhn, Thomas S. (1976), 'Theory-Change as Structure-Change: Comments on the Sneed Formalism'. *Erkenntnis* **10**, 179–99.

Kuhn, Thomas S. (1977), *The Essential Tension*. Chicago: University of Chicago Press.

Krausz, Michael (1984), 'Relativism and Foundationalism: Some Distinctions and Strategies'. *The Monist* **67**, 395–404.

Lakatos, Imre (1970), 'Falsification and the Methodology of Scientific Research Programmes'. Included in Lakatos and Musgrave, pp. 91–196.

Lakatos, Imre and Musgrave, Alan (eds.) (1970), *Criticism and the Growth of Knowledge*. Cambridge: Cambridge University Press.

Laudan, Larry (1976), 'Two Dogmas of Methodology'. *Philosophy of Science* **43**, 585–97.

Laudan, Larry (1977), *Progress and Its Problems*. Berkeley: Univ. of California Press.

Laudan, Larry (1980), 'Views of Progress: Separating the Pilgrims from the Rakes'. *Philosophy of the Social Sciences* **10**, 273–86.

Laudan, Larry (1981a), 'A Confutation of Convergent Realism'. *Philosophy of Science* **48**, 19–49.

Laudan, Larry (1981b), 'Problems, Truth, and Consistency'. *Studies in History and Philosophy of Science*, **13**, 73–80.

Lear, Jonathan (1982), 'Leaving the World Alone'. *Journal of Philosophy* **LXXIX**, 382–403.

Levin, Michael E. (1979), 'On Theory-Change and Meaning-Change'. *Philosophy of Science* **46**, 407–24.

Lugg, Andrew (1979), 'Laudan and the Problem-Solving Approach to Scientific Progress and Rationality'. *Philosophy of the Social Sciences* **9**, 466–474.

Mackie, J. L. (1977), *Ethics: Inventing Right and Wrong*. New York: Penguin books.

Mandelbaum, Maurice (1979), 'Subjective, Objective, and Conceptual Relativisms'. *The Monist* **62**, 403–428; reprinted in Meiland and Krausz, pp. 34–61.

Matthews, Eric (1983), 'Review of Hilary Putnam, *Reason, Truth and History*'. *Philosophical Books* **24**, 114–116.

McMullin, Ernan (1974), 'Logicality and Rationality: Comments on Toulmin's Theory of Science'. In Seeger, R.J., and Cohen, R.S. (eds.), *Philosophical Foundations of Science*. Dordrecht: D. Reidel Publishing Company, pp. 415–430.

McMullin, Ernan (1979), 'Discussion Review: Laudan's *Progress and Its Problems*'. *Philosophy of Science* **46**, 623–44.

Meiland, Jack W. (1973), 'Cognitive Relativism: Popper and the Argument from Language'. *Philosophical Forum* **4**, 406–421.

Meiland, Jack W. (1974), 'Kuhn, Scheffler, and Objectivity in Science'. *Philosophy of Science* **41**, 179–187.

Meiland, Jack W. (1977), 'Concepts of Relative Truth'. *The Monist* **60**, 568–582.

Meiland, Jack W. (1979), 'Is Protagorean Relativism Self-Refuting?' *Grazer Philosophische Studien* **9**, 51–68.

Meiland, Jack W. (1980), 'On the Paradox of Cognitive Relativism'. *Metaphilosophy* **11**, 115–126.

Meiland, Jack W. and Michael Krausz (eds.) (1982), *Relativism: Cognitive and Moral*. Notre Dame: University of Notre Dame Press.

Moberg, Dale W. (1979), 'Are There Rival, Incommensurable Theories?' *Philosophy of Science* **46**, 244–62.

Musgrave, Alan (1979), 'Problems with Progress'. *Synthese* **42**, 443–64.

Newton-Smith, W.H. (1981), *The Rationality of Science*. London: Routledge and Kegan Paul.

Nickles, Thomas (1980), 'Scientific Problems: Three Empiricist Models'. Included in Asquith and Giere, pp. 3–19.

Passmore, John A. (1961), *Philosophical Reasoning*. London: G. Duckworth.

Plato (1961), *Theaetetus*. Translated by F.M. Cornford, in Edith Hamilton and Huntington Cairns (eds.), *The Collected Dialogues of Plato*. Bollingen Series, Pantheon Books, pp. 845–919.

Popper, Karl (1966), *The Open Society and Its Enemies, Volume 2*, Fifth Edition Revised. Princeton: Princeton University Press.

Popper, Karl (1970), 'Normal Science and its Dangers'. In Lakatos and Musgrave, pp. 51–58.

Popper, Karl (1976), 'The Myth of the Framework'. In Eugene Freeman (ed.), *The Abdication of Philosophy: Philosophy and the Public Good*. LaSalle, Illinois: Open Court, pp. 23–48.

Putnam, Hilary (1977), 'Realism and Reason'. *Proceedings and Addresses of the American Philosophical Association* **50**, 483–498. Reprinted in Putnam, *Meaning and the Moral Sciences*, pp. 123–140.

Putnam, Hilary (1978), *Meaning and the Moral Sciences*. London: Routledge and Kegan Paul.

Putnam, Hilary (1979), 'Reflections on Goodman's *Ways of Worldmaking*'. *Journal of Philosophy* **LXXVI**, 603–618.

Putnam, Hilary (1981), *Reason, Truth and History*. Cambridge: Cambridge University Press.

Putnam, Hilary (1982), 'Why Reason Can't Be Naturalized'. *Synthese* **52**, 3–23.

Quine, Willard Van Orman (1960), *Word and Object*. Cambridge: M.I.T. Press.

Quine, Willard Van Orman (1975), 'On Empirically Equivalent Systems of the World'. *Erkenntnis* **9**, 313–328.

Quine, Willard Van Orman (1981a), 'On the Very Idea of a Third Dogma'. In Quine, *Theories and Things*, pp. 38–42.

Quine, Willard Van Orman (1981b), *Theories and Things*. Cambridge: Harvard University Press.

Quine, Willard Van Orman (1984), 'Relativism and Absolutism'. *The Monist* **67**, 293–6.

Rescher, Nicholas (1980), 'Conceptual Schemes'. In P.A. French, T.E. Uehling, Jr., and H.K. Wettstein (eds.), *Midwest Studies in Philosophy V*. Minneapolis: University of Minnesota Press, pp. 323–345.

Reichenbach, Hans (1938), *Experience and Prediction*. Chicago: University of Chicago Press.

Rorty, Richard (1972), The World Well Lost'. *Journal of Philosophy* **LXIX**, 649–665. Reprinted in Rorty, *Consequences of Pragmatism*, pp. 3–18.

Rorty, Richard (1979), *Philosophy and the Mirror of Nature*. Princeton: Princeton University Press.

Rorty, Richard (1982), *Consequences of Pragmatism*. Minneapolis: University of Minnesota Press.

Rorty, Richard (1983), 'Relativism'. Howison Lecture (typescript).

Scheffler, Israel (1963), *The Anatomy of Inquiry*. New York: Bobbs-Merrill.

Scheffler, Israel (1965), *Conditions of Knowledge*. Glenview, Illinois: Scott, Foresman and Company.

Scheffler, Israel (1972), 'Vision and Revolution: A Postscript on Kuhn'. *Philosophy of Science* **39**, 366–74.

Scheffler, Israel (1980), 'The Wonderful Worlds of Goodman'. *Synthese* **45**, 201–209.

Scheffler, Israel (1982), *Science and Subjectivity*. Second edition. Indianapolis: Hackett Publishing Company. First published in 1967.

Shapere, Dudley (1964), 'The Structure of Scientific Revolutions'. *Philosophical Review* **73**, 383–94.

Shapere, Dudley (1966), 'Meaning and Scientific Change'. In R.G. Colodny (ed.): *Mind and Cosmos*. Pittsburg: University of Pittsburg Press, pp. 41–85.

Shapere, Dudley (1971), 'The Paradigm Concept.' *Science* **172**, 706–9.

Shimony, Abner (1976), 'Comments on Two Epistemological Theses of Thomas Kuhn', in

R.S. Cohen *et al.* (eds.), *Essays in Memory of Imre Lakatos*. Boston Studies in the Philosophy of Science, Vol. 39. Dordrecht: D. Reidel Publishing Company, pp. 569–88.

Siegel, Harvey (1976), 'Meiland on Scheffler, Kuhn, and Objectivity in Science'. *Philosophy of Science* **43**, 441–8.

Siegel, Harvey (1980a), 'Justification, Discovery and the Naturalizing of Epistemology'. *Philosophy of Science* **47**, 297–321.

Siegel, Harvey (1980b), 'Objectivity, Rationality, Incommensurability, and More.' *British Journal for the Philosophy of Science* **31**, 359–375.

Siegel, Harvey (1980c), 'Epistemological Relativism in its Latest Form'. *Inquiry* **23**, 107–117.

Siegel, Harvey (1980d), 'Rationality, Talking Dogs, and Forms of Life'. *Educational Theory* **30**, 135–148.

Siegel, Harvey (1982a), 'On the Parallel Between Piagetian Cognitive Development and the History of Science'. *Philosophy of the Social Sciences* **12**, 375–386.

Siegel, Harvey (1982b), 'Relativism Refuted'. *Educational Philosophy and Theory* **14**, 47–50.

Siegel, Harvey (1983a), 'Truth, Problem Solving, and the Rationality of Science'. *Studies in History and Philosophy of Science* **14**, 89–112.

Siegel, Harvey (1983b), 'Brown on Epistemology and the New Philosophy of Science'. *Synthese* **56**, 61–89.

Siegel, Harvey (1984a), 'Empirical Psychology, Naturalized Epistemology, and First Philosophy'. *Philosophy of Science* **51**, 667–676.

Siegel, Harvey (1984b), 'Goodmanian Relativism'. *The Monist* **67**, 359–375.

Siegel, Harvey (1984c), 'Relativism, Realism and Rightness: Notes on Goodmanian Worldmaking'. *Journal of Thought* **19**, 16–35.

Siegel, Harvey (1985a), 'What Is the Question Concerning the Rationality of Science?' *Philosophy of Science* **52**, 517–537.

Siegel, Harvey (1985b), 'Relativism, Rationality, and Science Education'. *Journal of College Science Teaching* **15**, 102–105.

Siegel, Harvey (1985c), 'Reply to Elgin'. *Journal of Thought* **20**, 45–51.

Skagestad, Peter (1980), 'Pragmatic Realism: The Peircean Argument Reexamined'. *The Review of Metaphysics* **33**, 527–540.

Stegmüller, Wolfgang (1975), 'Structures and Dynamics of Theories: Some Reflections on J.D. Sneed and T.S. Kuhn', *Erkenntnis* **9**, 75–100.

Stegmüller, Wolfgang (1976a), *The Structure and Dynamics of Theories*. Translated from the German *Theorienstruckturen und Theoriendynamik* by Dr. William Wohlheuter. New York: Springer-Verlag.

Stegmüller, Wolfgang (1976b), 'Accidental ("Non-Substantial") Theory Change and Theory Dislodgement: To What Extent Logic Can Contribute to a Better Understanding of Certain Phenomena in the Dynamics of Theories'. *Erkenntnis* **10**, 147–78.

Strawson, P.F. (1982), 'Review of *Transcendental Arguments and Science*'. *Journal of Philosophy* **LXXIX**, 45–50.

Stroud, Barry (1969), 'Conventionalism and the Indeterminacy of Translation'. In Davidson and Hintikka, pp. 82–96.

Suppe, Frederick (1977), *The Structure of Scientific Theories*, 2nd edition. Urbana, Illinois: University of Illinois Press.

Swoyer, Chris (1982), 'True For'. In Meiland and Krausz, pp. 84–108.

Thagard, Paul (1980), 'Against Evolutionary Epistemology'. In Asquith and Giere, pp. 187–196.

Trigg, Roger (1973), *Reason and Commitment*. Cambridge: Cambridge University Press.

Toulmin, Stephen (1972), *Human Understanding*. Volume 1. Princeton: Princeton University Press.

Watkins, John W.N. (1970), 'Against Normal Science.' In Lakatos and Musgrave, pp. 25–37.

Weinert, Friedel (1984), 'Contra Res Sempiternas'. *The Monist* **67**, 376–394.

Whittaker, John H. (1978), 'Language-Games and Forms of Life Unconfused'. *Philosophical Investigations* **1**, 39–48.

Williams, Bernard (1972), 'An Inconsistent Form of Relativism'. From Williams, *Morality: An Introduction to Ethics* (New York: Harper and Row), pp. 22–26. Reprinted in Meiland and Krausz, pp. 171–174.

Williams, Bernard (1975), 'The Truth in Relativism'. *Proceedings of the Aristotelian Society* **75**, 215–228. Reprinted in Meiland and Krausz, pp. 175–185.

Winch, Peter (1958), *The Idea of a Social Science*. London: Routledge and Kegan Paul.

Wittgenstein, Ludwig (1953), *Philosophical Investigations*. New York: Macmillan Co.

Wong, David (1984), *Moral Relativity*. Berkeley: University of California Press.

Wykstra, Stephen J. (1980), 'Toward a Historical Meta-Method for Assessing Normative Methodologies: Rationality, Serendipity, and the Robinson Crusoe Fallacy'. In Asquith and Giere, pp. 211–222.

INDEX

SYNTHESE LIBRARY

Studies in Epistemology, Logic, Methodology,
and Philosophy of Science

1. J. M. Bochénski, *A Precis of Mathematical Logic.* 1959.
2. P. L. Guiraud, *Problèmes et méthodes de la statistique linguistique.* 1960.
3. Hans Freudenthal (ed.), *The Concept and the Role of the Model in Mathematics and Natural and Social Sciences.* 1961.
4. Evert W. Beth, *Formal Methods. An Introduction to Symbolic Logic and the Study of Effective Operations in Arithmetic and Logic.* 1962.
5. B. H. Kazemier and D. Vuysje (eds.), *Logic and Language. Studies Dedicated to Professor Rudolf Carnap on the Occasion of His Seventieth Birthday.* 1962.
6. Marx W. Wartofsky (ed.), *Proceedings of the Boston Colloquium for the Philosophy of Science 1961–1962.* Boston Studies in the Philosophy of Science, Volume I. 1963.
7. A. A. Zinov'ev, *Philosophical Problems of Many-Valued Logic.* 1963.
8. Georges Gurvitch, *The Spectrum of Social Time.* 1964.
9. Paul Lorenzen, *Formal Logic.* 1965.
10. Robert S. Cohen and Marx W. Wartofsky (eds.), *In Honor of Philipp Frank.* Boston Studies in the Philosophy of Science, Volume II. 1965.
11. Evert W. Beth, *Mathematical Thought. An Introduction to the Philosophy of Mathematics.* 1965.
12. Evert W. Beth and Jean Piaget, *Mathematical Epistemology and Psychology.* 1966.
13. Guido Küng, *Ontology and the Logistic Analysis of Language. An Enquiry into the Contemporary Views on Universals.* 1967.
14. Robert S. Cohen and Marx W. Wartofsky (eds.), *Proceedings of the Boston Colloquium for the Philosophy of Sciences 1964–1966. In Memory of Norwood Russell Hanson.* Boston Studies in the Philosophy of Science, Volume III. 1967.
15. C. D. Broad, *Induction, Probability, and Causation. Selected Papers.* 1968.
16. Günther Patzig, *Aristotle's Theory of the Syllogism. A Logical-Philosophical Study of Book A of the Prior Analytics.* 1968.
17. Nicholas Rescher, *Topics in Philosophical Logic.* 1968.
18. Robert S. Cohen and Marx W. Wartofsky (eds.), *Proceedings of the Boston Colloquium for the Philosophy of Science 1966–1968.* Boston Studies in the Philosophy of Science, Volume IV. 1969

19. Robert S. Cohen and Marx W. Wartofsky (eds.), *Proceedings of the Boston Colloquium for the Philosophy of Science 1966*968. Boston Studies in the Philosophy of Science, Volume V. 1969
20. J. W. Davis, D. J. Hockney, and W. K. Wilson (eds.), *Philosophical Logic*. 1969
21. D. Davidson and J. Hintikka (eds.), *Words and Objections. Essays on the Work of W. V. Quine*. 1969.
22. Patrick Suppes. *Studies in the Methodology and Foundations of Science. Selected Papers from 1911 to 1969*. 1969
23. Jaakko Hintikka, *Models for Modalities. Selected Essays*. 1969
24. Nicholas Rescher *et al.* (eds.), *Essays in Honor of Carl G. Hempel. A Tribute on the Occasion of His Sixty-Fifth Birthday*. 1969
25. P. V. Tavanec (ed.), *Problems of the Logic of Scientific Knowledge*. 1969
26. Marshall Swain (ed.), *Induction, Acceptance, and Rational Belief*. 1970.
27. Robert S. Cohen and Raymond J. Seeger (eds.), *Ernst Mach: Physicist and Philosopher*. Boston Studies in the Philosophy of Science, Volume V. 1970.
28. Jaakko Hintikka and Patrick Suppes, *Information and Inference*. 1970.
29. Karel Lambert, *Philosophical Problems in Logic. Some Recent Developments*. 1970.
30. Rolf A. Eberle, *Nominalistic Systems*. 1970.
31. Paul Weingartner and Gerhard Zecha (eds.), *Induction, Physics, and Ethics*. 1970.
32. Evert W. Beth, *Aspects of Modern Logic*. 1970.
33. Risto Hilpinen (ed.), *Deontic Logic: Introductory and Systematic Readings*. 1971.
34. Jean-Louis Krivine, *Introduction to Axiomatic Set Theory*. 1971.
35. Joseph D. Sneed, *The Logical Structure of Mathematical Physics*. 1971.
36. Carl R. Kordig, *The Justification of Scienntific Change*. 1971.
37. Milic Capek, *Bergson and Modern Physics*. Boston Studies in the Philosophy of Science, Volume VII. 1971.
38. Norwood Russell Hanson, *What I Do Not Believe, and Other Essays* (ed. by Stephen Toulmin and Harry Woolf). 1971.
39. Roger C. Buck and Robert S. Cohen (eds.), *PSA 1970. In Memory of Rudolf Carnap*. Boston Studies in the Philosophy of Science, Volume VIII. 1971
40. Donald Davidson and Gilbert Harman (eds.), *Semantics of Natural Language*. 1972.
41. Yehoshua Bar-Hillel (ed.), *Pragmatics of Natural Languages*. 1971.
42. Sören Stenlund, *Combinators, λ-Terms and Proof Theory*. 1972.
43. Martin Strauss, *Modern Physics and Its Philosophy. Selected Paper in the Logic, History, and Philosophy of Science*. 1972.
44. Mario Bunge, *Method, Model and Matter*. 1973.
45. Mario Bunge, *Philosophy of Physics*. 1973.
46. A. A. Zinov'ev, *Foundations of the Logical Theory of Scientific Knowledge (Complex Logic)*. (Revised and enlarged English edition with an appendix by G. A. Smirnov, E. A. Sidorenka, A. M. Fedina, and L. A. Bobrova.) Boston Studies in the Philosophy of Science, Volume IX. 1973.
47. Ladislav Tondl, *Scientific Procedures*. Boston Studies in the Philosophy of Science, Volume X. 1973.
48. Norwood Russell Hanson, *Constellations and Conjectures* (ed. by Willard C. Humphreys, Jr.). 1973
49. K. J. J. Hintikka, J. M. E. Moravcsik, and P. Suppes (eds.), *Approaches to Natural Language*. 1973.
50. Mario Bunge (ed.), *Exact Philosophy – Problems, Tools, and Goals*. 1973.

51. Radu J. Bogdan and Ilkka Niiniluoto (eds.), *Logic, Language, and Probability*. 1973.
52. Glenn Pearce and Patrick Maynard (eds.), *Conceptual Change*. 1973.
53. Ilkka Niiniluoto and Raimo Tuomela, *Theoretical Concepts and Hypothetico-Inductive Inference*. 1973.
54. Roland Fraissé, *Course of Mathematical Logic* – Volume 1: *Relation and Logical Formula*. 1973.
55. Adolf Grünbaum, *Philosophical Problems of Space and Time*. (Second, enlarged edition.) Boston Studies in the Philosophy of Science, Volume XII. 1973.
56. Patrick Suppes (ed.), *Space, Time, and Geometry*. 1973.
57. Hans Kelsen, *Essays in Legal and Moral Philosophy* (selected and introduced by Ota Weinberger). 1973.
58. R. J. Seeger and Robert S. Cohen (eds.), *Philosophical Foundations of Science*. Volume XI. 1974.
59. Robert S. Cohen and Marx W. Wartofsky (eds.), *Logical and Epistemological Studies in Contemporary Physics*. Boston Studies in the Philosophy of Science, Volume XIII. 1973.
60. Robert S. Cohen and Marx W. Wartofsky (eds.), *Methodological and Historical Essays in the Natural and Social Sciences. Proceedings of the Boston Colloquium for the Philosophy of Science 1969–1972*. Boston Studies in the Philosophy of Science, Volume XIV. 1974.
61. Robert S. Cohen, J. J. Stachel, and Marx W. Wartofsky (eds.), *For Dirk Struik. Scientific, Historical and Political Essays in Honor of Dirk J. Struik*. Boston Studies in the Philosophy of Science, Volume XV. 1974.
62. Kazimierz Ajdukiewicz, *Pragmatic Logic* (transl. from the Polish by Olgierd Wojtasiewicz). 1974.
63. Sören Stenlund (ed.), *Logical Theory and Semantic Analysis. Essays Dedicated to Stig Kanger on His Fiftieth Birthday*. 1974.
64. Kenneth F. Schaffner and Robert S. Cohen (eds.), *Proceedings of the 1972 Biennial Meeting, Philosophy of Science Association*. Boston Studies in the Philosophy of Science, Volume XX. 1974.
65. Henry E. Kyburg, Jr., *The Logical Foundations of Statistical Inference*. 1974.
66. Marjorie Grene, *The Understanding of Nature. Essays in the Philosophy of Biology*. Boston Studies in the Philosophy of Science, Volume XXIII. 1974.
67. Jan M. Broekman, *Structuralism: Moscow, Prague, Paris*. 1974.
68. Norman Geschwind, *Selected Papers on Language and the Brain*, Boston Studies in the Philosophy of Science, Volume XVI. 1974.
69. Roland Fraissé, *Course of Mathematical Logic* – Volume 2: *Model Theory*. 1974.
70. Andrzej Grzegorczyk, *An Outline of Mathematical Logic. Fundamental Results and Notions Explained with All Details*. 1974.
71. Franz von Kutschera, *Philosophy of Language*. 1975.
72. Juha Manninen and Raimo Tuomela (eds.), *Essays on Explanation and Understanding. Studies in the Foundations of Humanities and Social Sciences*. 1976.
73. Jaakko Hintikka (ed.), *Rudolf Carnap, Logical Empiricist. Materials and Perspectives*. 1975.
74. Milic Capek (ed.), *The Concepts of Space and Time. Their Structure and Their Development*. Boston Studies in the Philosophy of Science, Volume XXII. 1976.
75. Jaakko Hintikka and Unto Remes, *The Method of Analysis. Its Geometrical Origin and Its General Significance*. Boston Studies in the Philosophy of Science, Volume XXV. 1974.

76. John Emery Murdoch and Edith Dudley Sylla, *The Cultural Context of Medieval Learning*. Boston Studies in the Philosophy of Science, Volume XXVI. 1975.
77. Stefan Amsterdamski, *Between Experience and Metaphysics. Philosophical Problems of the Evolution of Science*. Boston Studies in the Philosophy of Science, Volume XXXV. 1975.
78. Patrick Suppes (ed.), *Logic and Probability in Quantum Mechanics*. 1976.
79. Hermann von Helmholtz: *Epistemological Writings. The Paul Hertz/Moritz Schlick Centenary Edition of 1921 with Notes and Commentary by the Editors*. (Newly translated by Malcolm F. Lowe. Edited, with an Introduction and Bibliography, by Robert S. Cohen and Yehuda Elkana.) Boston Studies in the Philosophy of Science, Volume XXVIII. 1975.
80. Joseph Agassi, *Science in Flux*. Boston Studies in the Philosophy of Science, Volume XXVIII. 1975.
81. Sandra G. Harding (ed.), *Can Theories Be Refuted? Essays on the Duhem-Quine Thesis*. 1976.
82. Stefan Nowak, *Methodology of Sociological Research. General Problems*. 1977.
83. Jean Piaget, Jean-Blaise Grize, Alina Szeminska, and Vinh Bang, *Epistemology and Psychology of Functions*. 1977.
84. Marjorie Grene and Everett Mendelsohn (eds.), *Topics in the Philosophy of Biology*. Boston Studies in the Philosophy of Science, Volume XXVII. 1976.
85. E. Fischbein, *The Intuitive Sources of Probabilistic Thinking in Children*. 1975.
86. Ernest W. Adams, *The Logic of Conditionals. An Application of Probability to Deductive Logic*. 1975.
87. Marian Przelecki and Ryszard Wójcicki (eds.), *Twenty-Five Years of Logical Methodology in Poland*. 1976.
88. J. Topolski, *The Methodology of History*. 1976.
89. A. Kasher (ed.), *Language in Focus: Foundations, Methods and Systems. Essays Dedicated to Yehoshua Bar-Hillel*. Boston Studies in the Philosophy of Science, Volume XLIII. 1976.
90. Jaakko Hintikka, *The Intentions of Intentionality and Other New Models for Modalities*. 1975.
91. Wolfgang Stegmüller, *Collected Papers on Epistemology, Philosophy of Science and History of Philosophy*. 2 Volumes. 1977.
92. Dov M. Gabbay, *Investigations in Modal and Tense Logics with Applications to Problems in Philosophy and Linguistics*. 1976.
93. Radu J. Bodgan, *Local Induction*. 1976.
94. Stefan Nowak, *Understanding and Prediction. Essays in the Methodology of Social and Behavioral Theories*. 1976.
95. Peter Mittelstaedt, *Philosophical Problems of Modern Physics*. Boston Studies in the Philosophy of Science, Volume XVIII. 1976.
96. Gerald Holton and William Blanpied (eds.), *Science and Its Public: The Changing Relationship*. Boston Studies in the Philosophy of Science, Volume XXXIII. 1976.
97. Myles Brand and Douglas Walton (eds.), *Action Theory*. 1976.
98. Paul Gochet, *Outline of a Nominalist Theory of Proposition. An Essay in the Theory of Meaning*. 1980.
99. R. S. Cohen, P. K. Feyerabend, and M. W. Wartofsky (eds.), *Essays in Memory of Imre Lakatos*. Boston Studies in the Philosophy of Science, Volume XXXIX. 1976.
100. R. S. Cohen and J. J. Stachel (eds.), *Selected Papers of Léon Rosenfield*. Boston Studies in the Philosophy of Science, Volume XXI. 1978.

101. R. S. Cohen, C. A. Hooker, A. C. Michalos, and J. W. van Evra (eds.), *PSA 1974: Proceedings of the 1974 Biennial Meeting of the Philosophy of Science Association*. Boston Studies in the Philosophy of Science, Volume XXXII. 1976.
102. Yehuda Fried and Joseph Agassi, *Paranoia: A Study in Diagnosis*. Boston Studies in the Philosophy of Science, Volume L. 1976.
103. Marian Przelecki, Klemens Szaniawski, and Ryszard Wójcicki (eds.), *Formal Methods in the Methodology of Empirical Sciences*. 1976.
104. John M. Vickers, *Belief and Probability*. 1976.
105. Kurt H. Wolff, *Surrender and Catch: Experience and Inquiry Today*. Boston Studies in the Philosophy of Science, Volume LI. 1976.
106. Karel Kosík, *Dialectics of the Concrete*. Boston Studies in the Philosophy of Science, Volume LII. 1976.
107. Nelson Goodman, *The Structure of Appearance* (Third edition.) Boston Studies in the Philosophy of Science, Volume LIII. 1977.
108. Jerzy Giedymin (ed.), *Kazimierz Ajdukiewicz: The Scientific World-Perspective and Other Essays, 1931-1963*. 1978.
109. Robert L. Causey, *Unity of Science*. 1977.
110. Richard E. Grandy, *Advanced Logic for Applications*. 1977.
111. Robert P. McArthur, *Tense Logic*. 1976.
112. Lars Lindahl, *Position and Change. A Study in Law and Logic*. 1977.
113. Raimo Tuomela, *Dispositions*. 1978.
114. Herbert A. Simon, *Models of Discovery and Other Topics in the Methods of Science*. Boston Studies in the Philosophy of Science, Volume LIV. 1977.
115. Roger D. Rosenkrantz, *Inferenc, Method and Decision*. 1977.
116. Raimo Tuomela, *Human Action and Its Explanation. A Study on the Philosophical Foundations of Psychology*. 1977.
117. Morris Lazerowitz, *The Language of Philosophy. Freud and Wittgenstein*. Boston Studies in the Philosophy of Science, Volume LV. 1977.
118. Stanislaw Leśniewski, *Collected Works* (ed. by S. J. Surma, J. T. J. Srzednicki, and D. I. Barnett, with an annotated bibliography by V. Frederick Rickey). 1982. (Forthcoming.)
119. Jerzy Pelc, *Semiotics in Poland, 1894-1969*. 1978.
120. Ingmar Pörn, *Action Theory and Social Science. Some Formal Models*. 1977.
121. Joseph Margolis, *Persons and Mind. The Prospects of Nonreductive Materialism*. Boston Studies in the Philosophy of Science, Volume LVII. 1977.
122. Jaakko Hintikka, Ilkka Niiniluoto, and Esa Saarinen (eds.), *Essays on Mathematical and Philosophical Logic*. 1978.
123. Theo A. F. Kuipers, *Studies in Inductive Probability and Rational Expectation*. 1978.
124. Esa Saarinen, Risto Hilpinen, Ilkka Niiniluoto, and Merrill Provence Hintikka (eds.), *Essays in Honour of Jaakko Hintikka on the Occasion of His Fiftieth Birthday*. 1978.
125. Gerard Radnitzky and Gunnar Andersson (eds.),*Progress and Rationality in Science*. Boston Studies in the Philosophy of Science, Volume LVIII. 1978.
126. Peter Mittelstaedt, *Quantum Logic*. 1978.
127. Kenneth A. Bowen, *Model Theory for Modal Logic. Kripke Models for Modal Predicate Calculi*. 1978.
128. Howard Alexander Bursen, *Dismantling the Memory Machine. A Philosophical Investigation of Machine Theories of Memory*. 1978.
129. Marx W. Wartofsky, *Models: Representation and the Scientific Understanding*. Boston Studies in the Philosophy of Science, Volume XLVIII. 1979.

130. Don Ihde, *Technics and Praxis. A Philosophy of Technology*. Boston Studies in the Philosophy of Science, Volume XXIV. 1978.
131. Jerzy J. Wiatr (ed.), *Polish Essays in the Methodology of the Social Sciences.* Boston Studies in the Philosophy of Science, Volume XXIX. 1979.
132. Wesley C. Salmon (ed.), *Hans Reichenbach: Logical Empiricist*. 1979.
133. Peter Bieri, Rolf-P. Horstmann, and Lorenz Krüger (eds.), *Transcendental Arguments in Science Essays in Epistemology*. 1979.
134. Mihailo Marković and Gajo Petrović (eds.), *Praxis, Yugoslav Essays in the Philosophy and Methodology of the Social Sciences*. Boston Studies in the Philosophy of Science, Volume XXXVI. 1979.
135. Ryszard Wójcicki, *Topics in the Formal Methodology of Empirical Sciences*. 1979.
136. Gerard Radnitzky and Gunnar Andersson (eds.), *The Structure and Development of Science*. Boston Studies in the Philosophy of Science, Volume LIX. 1979.
137. Judson Chambers Webb, *Mechanism, Mentalism, and Metamathematics. An Essay on Finitism*. 1980.
138. D. F. Gustafson and B. L. Tapscott (eds.), *Body, Mind, and Method. Essays in Honor of Virgil C. Aldrich*. 1979.
139. Leszek Nowak, *The Structure of Idealization. Towards a Systematic Interpretation of the Marxian Idea of Science*. 1979.
140. Chaim Perelman, *The New Rhetoric and the Humanities. Essays on Rhetoric and Its Applications*. 1979.
141. Wlodzimierz Rabinowicz, *Universalizability. A Study in Morals and Metaphysics*. 1979.
142. Chaim Perelman, *Justice, Law, and Argument. Essays on Moral and Legal Reasoning*. 1980.
143. Stig Kanger and Sven Öhman (eds.), *Philosophy and Grammar. Papers on the Occasion of the Quincentennial of Uppsala University*.
144. Tadeusz Pawlowski, *Concept Formation in the Humanities and the Social Sciences*. 1980.
145. Jaakko Hintikka, David Gruender, and Evandro Agazzi (eds.), *Theory Change, Ancient Axiomatics, and Galileo's Methodology. Proceedings of the 1978 Pisa Conference on the History and Philosophy of Science,* Volume I. 1981.
146. Jaakko Hintikka, David Gruender, and Evandro Agazzi *Probabilistic Thinking, Thermodynamics, and the Interaction of the History and Philosophy of Science. Proceedings of the 1978 Pisa Conference on the History and Philosophy of Science,* Volume II. 1981.
147. Uwe Mönnich (ed.), *Aspects of Philosophical Logic. Some Logical Forays into Central Notions of Linguistics and Philosophy*. 1981.
148. Dov M. Gabbay, *Semantical Investigations in Heyting's Intuitionistic Logic*. 1981.
149. Evandro Agazzi (ed.), *Modern Logic – A Survey. Historical, Philosophical, and Mathematical Aspects of Modern Logic and its Applications*. 1981.
150. A. F. Parker-Rhodes, *The Theory of Indistinguishables. A Search for Explanatory Principles below the Level of Physics*. 1981.
151. J. C. Pitt, *Pictures, Images, and Conceptual Change. An Analysis of Wilfrid Sellars' Philosophy of Science*. 1981.
152. R. Hilpinen (ed.), *New Studies in Deontic Logic. Norms, Actions, and the Foundations of Ethics*. 1981.
153. C. Dilworth, *Scientific Progress. A Study Concerning the Nature of the Relation Between Successive Scientific Theories*. 1981.
154. D. W. Smith and R. McIntyre, *Husserl and Intentionality. A Study of Mind, Meaning, and Language*. 1982.

155. R. J. Nelson, *The Logic of Mind.* 1982.
156. J. F. A. K. van Benthem, *The Logic of Time. A Model-Theoretic Investigation into the Varieties of Temporal Ontology, and Temporal Discourse.* 1982.
157. R. Swinburne (ed.), *Space, Time and Causality.* 1982.
158. R. D. Rozenkrantz, *E. T. Jaynes: Papers on Probability, Statistics and Statistical Physics.* 1983.
159. T. Chapman, *Time: A Philosophical Analysis.* 1982.
160. E. N. Zalta, *Abstract Objects. An Introduction to Axiomatic Metaphysics.* 1983.
161. S. Harding and M. B. Hintikka (eds.), *Discovering Reality. Feminist Perspectives on Epistemology, Metaphysics, Methodology, and Philosophy of Science.* 1983.
162. M. A. Stewart (ed.), *Law, Morality and Rights.* 1983.
163. D. Mayr and G. Süssmann (eds.), *Space, Time, and Mechanics. Basic Structure of a Physical Theory.* 1983.
164. D. Gabbay and F. Guenthner (eds.), *Handbook of Philosophical Logic.* Vol. I. 1983.
165. D. Gabbay and F. Guenthner (eds.), *Handbook of Philosophical Logic.* Vol. II. 1984.
166. D. Gabbay and F. Guenthner (eds.), *Handbook of Philosophical Logic.* Vol. III. 1985.
167. D. Gabbay and F. Guenthner (eds.), *Handbook of Philosophical Logic.* Vol. IV, forthcoming.
168. Andrew, J. I. Jones, *Communication and Meaning.* 1983.
169. Melvin Fitting, *Proof Methods for Modal and Intuitionistic Logics.* 1983.
170. Joseph Margolis, *Culture and Cultural Entities.* 1984.
171. Raimo Tuomela, *A Theory of Social Action.* 1984.
172. Jorge J. E. Gracia, Eduardo Rabossi, Enrique Villanueva, and Marcelo Dascal (eds.), *Philosophical Analysis in Latin America.* 1984.
173. Paul Ziff, *Epistemic Analysis. A Coherence Theory of Knowledge.* 1984.
174. Paul ZIff, *Antiaesthetics. An Appreciation of the Cow with the Subtile Nose.* 1984.
175. Wolfgang Balzer, David A. Pearce, and Heinz-Jürgen Schmidt (eds.), *Reduction in Science. Structure, Examples, Philosophical Problems.* 1984.
176. Aleksander Peczenik, Lars Lindahl, and Bert van Roermund (eds.), *Theory of Legal Science. Proceedings of the Conference on Legal Theory and Philosophy of Science, Lund, Sweden, December 11–14, 1983.* 1984.
177. Ilkka Niiniluoto, *Is Science Progressive?* 1984.
178. Binal Matilal and Jaysankar Lal Shaw (eds.), *Exploratory Essays in Current Theories and Classical Indian Theories of Meaning and Reference.* 1985.
179. Peter Kroes, *Time: Its Structure and Role in Physical Theories.* 1985.
180. James H. Fetzer, *Sociobiology and Epistemology,* 1985.
181. L. Haaparanta and J. Hintikka, *Frege Synthesized. Essays on the Philosophical and Foundational Work of Gottlob Frege.* 1986.
182. Michael Detlefsen, *Hilbert's Program. An Essay on Mathematical Instrumentalism.* 1986.